PHYSICS RESEARCH AND TECHNOLOGY

GENERAL RELATIVITY AND THE PIONEERS ANOMALY

PHYSICS RESEARCH AND TECHNOLOGY

Additional books in this series can be found on Nova's website
under the Series tab.

Additional E-books in this series can be found on Nova's website
under the E-books tab.

MATHEMATICS RESEARCH DEVELOPMENTS

Additional books in this series can be found on Nova's website
under the Series tab.

Additional E-books in this series can be found on Nova's website
under the E-books tab.

PHYSICS RESEARCH AND TECHNOLOGY

GENERAL RELATIVITY AND THE PIONEERS ANOMALY

MARCELO SAMUEL BERMAN

Nova Science Publishers, Inc.
New York

For permission to use material from this book please contact us:
Telephone 631-231-7269; Fax 631-231-8175
Web Site: http://www.novapublishers.com

Library of Congress Cataloging-in-Publication Data

General relativity and the pioneers anomaly / editor, Marcelo Samuel Berman.
p. cm.
Includes bibliographical references and index.
ISBN 978-1-62100-003-7 (hardcover)
1. General relativity (Physics) 2. Acceleration (Mechanics) 3. Relativistic quantum theory. I. Berman, Marcelo Samuel.
QC173.6.G4675 2011
530.11--dc23

2011029486

Published by Nova Science Publishers, Inc. † New York

Contents

For Albert, Álvaro, Clarissa, Elizabeth, Gabriel, Gustavo, Miguelina, Paula, Renata and Rosa

In honour of Solange Lima Kaczyk, my inspiring muse.

"Happy is the man that findeth wisdom, and the man that getteth understanding: for the merchandise of it, is better than the merchandise of silver, and the gain thereof than fine gold"... "Take fast hold of instruction; let her not go: keep her, for she is thy life".

(The Book of Proverbs, The Bible. Copyrighted by our DAD)

Preface and Introduction

This textbook is a simple introduction to General Relativity Theory (GRT), and to the solution of the Pioneers Anomaly, by means of Relativistic Cosmology, designed to be also understood,by senior undergraduates, or first year graduate students in any of the fields of Theoretical Physics,Applied Mathematics,or Space Engineering. It is the first elementary account of GRT, and Cosmology, to address the NASA problem, which consists in a specific deceleration suffered by two space-probes launched to outer space more than thirty years ago. In 2007, I gave a possible explanation, employing a semi-relativistic Machian approach, which led to admitting that the Universe has a peculiar kind of rotation,and that each observer in the Universe, sees any other observed point as being centripetally decelerated towards himself. The calculation I had made,coincided numerically, with the observed Pioneers anomalous deceleration.(Berman,2007). Recently, in collaboration with my former advisor, Professor Fernando de Mello Gomide, we were successful in demonstrating that the same solution can be found within General Relativity.(Berman and Gomide,2010;2011;-a). We feel like having performed Einstein´s dream, that the Universe could be understood if one had just paper and pencil. At the same time, I obtained other simplified proofs, like the ones on the zero-total energy and zero-total energy density of the Universe,(that are due to the negative contribution of the gravitational field), and the solution for the cosmological frame-dragging, plus a model where the total spin of the Universe becomes constant.I think that conservation of spin is a major achievement, since nobody ever declared that the Newtonian Theorem of angular momentum conservation would apply for the General Relativistic rotating and expanding Universe (Section 9.5).It also applies when convenient conditions are imposed in variable speed of light relativistic theory.

Should one prefer scientific theories that contain beautiful equations, rather than ugly ones? I am against prejudices, but beautiful equations are fundamental!!!If one adds the "simplicity" requirement, we are in paradise.A mathematical detail in the general relativistic formalism, forces us, in some models, to choose a negative angular speed and discard the positive one, showing a preference, that John D. Barrow was led to call the left-handed Universe.Also is very important–that there is no center of rotation, not even an axis, in its common interpretation..It is a rotation of the tri-space around the temporal "axis", which is orthogonal to the tri-space.And both positive or negative angular speed solutions, solve the Pioneers anomalous deceleration.

There is a secondary Pioneers Anomaly, that has been pointed out: the two spacecrafts, are slowing down their spins, except when there are disturbances, or the power engines are turned on.I offer an explanation, within the solution of the main anomaly, to solve the secondary one, through the rotation of the Universe.I warn everybody that my model predicts an angular acceleration of the Universe, of about 10^{-35} rad/s^2,for the present day, and in general,it is given by $\frac{cH}{R}$, i.e. the product of the speed of light and the Hubble´s parameter, divided by the radius of the causally related Universe. This angular acceleration frame-draggs locally the Universe, and you obtain the local angular accelerations, by replacing R by the local linear dimensions. For instance, the angular acceleration of the Pioneers spin is $\frac{cH}{r}$, where $r = 5$ meters , is the radius of the spacecraft.We are in face of the Einstein-Machian dream of a cosmological frame-dragging.

As I am a Brazilian Jew, and also an Argentine, I belong to both Brazilian minorities; when I was a 22 years old electronics engineer, teaching Physics, I was the youngest teacher, in any Latin American university, and then, the Brazilian dictatorial military regime, sponsored hatred and prejudice against me, ultimately causing an interruption in my career, for more than a decade. If such a disgrace had not happened to me, this textbook would have probably been written with more maturity. For many years, I have been waiting for a reparation. I appeal to Dilma Vana Rosseff,President of Brazil, to demand a final decision, in the Ministry of Justice.

The acceleration of the Universe, and the variation of the fine-structure "constant" are here given some space.The variable speed of light, is shown here to agree with our explanation of the possible rotation of the Universe and the Pioneers anomaly, due to the fact that there is an equivalence among those theories. The news accounting for the 2006 experimental evidence that the Universe has been $\Lambda-$ driven for a long period in its history, is not a surprise to me; this fact was evident from my treatments on the Machian Universe, where it was derived, from the zero-total energy of the Universe conjecture, that each particular form of energy density, would keep its relative importance towards the total energy density, during all the Universe's life. On the other hand, I have been particularly productive, during the late 80's and the early 90's, by publishing papers dealing with B.D. theory and lambda models, when almost no one attached importance to such problems. I remember that I had been warned by several colleagues, that I should not pursue such endeavors, "because no journal was interested" in these subjects.

The first two chapters of this book (Part I), deal with the mathematical foundation of GRT; just the bare rudiments of Riemannian geometry and tensor calculus that are needed for understanding this book are touched upon. Next we deal with GRT leading to Einstein's field equations and then, the Schwarzschild solution that represents a black hole or the solar system (Part II). It is shown, as a matter of fact, the fact that matter curves spacetime. Further chapters introduce the minimum possible account of Relativistic Cosmology (Part III), in order to deal with our goal which is the solution of cosmological models that predict a decelerating field like the one faced by the Pioneers(Part IV).Chapter 11 and 12 are diminute, and basically contain final comments.

The new century´s first decade, has brought, three important discoveries: the first, is the confirmation of the Lense-Thirring metric effects, reported by Ignazio Ciufolini and collaborators (2004), and the first general relativistic treatment of the rotational speeds in galaxies, by Fred Cooperstock and S. Tieu (2005), showing that the theory needs not be reinforced by inclusion of dark matter. The Lense-Thirring metric and other developments, as well as Cooperstock's approach to dark matter and rotational speeds in galaxies, are dealt in Berman (2007). The third one is the experimental discovery that our Universe has been lambda-driven for the most part of its history, and even in the present. Most of the treatment given to cosmological models in this book, preserve these observational results. The cosmological constant problem, as we hope, will soon not be a constant problem in our life. This book announces new confirmations of General Relativity, through the discovery of the Pioneers Anomaly. It is at the same time a confirmation of the rotation of the Universe, cosmological frame-dragging,and of the existence of cosmological gravitomagnetism, in my opinion.

My thanks are due to Professor Fernando de Mello Gomide, who coauthored two books with me (Berman and Gomide, 1987; Gomide and Berman, 1988), parts of which are included in the present one. I thank the deceased Professor Murari Mohan Som for the advices given by him to me during the last twenty years. Both Prof. Gomide and Prof. Som were my advisors and intellectual mentors and afterwards, friends and colleagues; in place of Prof. Som, now I have had the friendship of Prof. Newton C.A. da Costa, the father of modern mathematical logic in Latin America,and a specialist in the foundations of Science. He even was my former high school teacher of mathematics, and always gave me a lot of his most friendly incentive. Henrique Fleming was also important in my doctoral thesis examination.

I remain grateful to Professors Fred Cooperstock, James Ipser and S. Gottesman, and also to the friendly colleagues I had during the several months I stayed at the University of Alabama. Maurice Bazin, one of the two authors that are alive, out of the three (Adler, Bazin and M. Schiffer) who wrote the unsurpassed introductory book on General Relativity, was gentle enough in many occasions. My public thanks go also to Lucasian Professor Stephen Hawking and his staff, who allowed me to prepare and edit the Brazilian version of "*The Illustrated a Brief History of Time*".

Most of the usual cosmological models employed by cosmologists, are of the constant deceleration parameter type. I have the credit for having derived them (Berman, 1983; Berman and Gomide, 1988), when no one had called the attention to the fact that the only two particular cases that were known (Rowan-Robinson, 1981), pointed out to these generalized models.. This textbook deals with mathematical cosmological models, not with experimental Cosmology. Being this an openly didactic book, we have inserted several repetitive passages which, we hope, will not bore the more competent readers. Repetition is necessary if you want to instruct beginning students. Of course, you would not be so redundant in a doctoral thesis.

We have sacrificed mathematical rigor, in order that readers with different backgrounds, would not become exaggeratedly frustrated. We change freely from Latin to Greek indices and vice-versa, without penalty: the reader shall easily decide in each particular case if the index runs from 0 to 3 or from 1 to 3. At times, we suppose that the constants G and

c have unit value; in other times, we include these constants into the text. Later, while reading scientific papers on the subject, some difficulties of this kind will, then, be easily overcome. The pre-requisites for this book are University Physics (at the level of Halliday et al., 2005; Ohanian, 1985), and Advanced Calculus (Wrede and Spiegel, 2002).The sections on energy-momentum pseudotensors, can be skipped by the readers, if they are not familiar with the formalism.

I owe a special gratitude debt with Mr. Marcelo F. Guimarães who has collaborated with TEX, throughout this text and many others. We are not native English speakers, so I ask the reader's indulgence with the eventual and unavoidable language mistakes.

Maybe I am wrong with the rotation of the Universe. I take the sole responsibility for the errors that may be in the text. I ask the reader to blame me for anything he does not like; but he should praise Professors F.M. Gomide, Newton da Costa,and the late M.M. Som in the opposite case.

Last, but not least, the author extends his thanks, to Frank Columbus, former President, and Nadya Columbus,President of Nova Science Publishers, and its staff, for allowing me to publish books, and to the important role they had,in all the phases of the production, and to them I give part of the credit for this book.

Prof. Marcelo Samuel Berman
Instituto Albert Einstein/Latinamerica
msberman@institutoalberteinstein.org
Curitiba - June, 2011.

Part I

Mathematical Preliminaries

Chapter 1

Tensors in Amorphous Spaces

1.1. Introduction

Let $x^1, x^2, ..., x^N$ be N variables. Each set of N possible values for these variables is called a point. The variables are called coordinates, and the totality of points corresponding to all the possible values of the coordinates, defines an N - dimensional space. Other names usually given are "manifolds", or "hyperspaces", which means basically the same thing. The particular case $N = 4$ is usually well known for its rôle in the Theory of Relativity, where the four dimensional space is called spacetime.

The Tensor Calculus is a tool which establishes a mathematical language valid for any system of coordinates, and not only in a particular one. As it will become clear later, equations written with the help of tensors benefit from this property. Nobel prize winner Albert Einstein made recourse to this tool (Tensor Calculus) in order to establish relations valid for any observers in arbitrary relative motion, by stating the Principle of Covariance of the physical laws. It is very evident that a given physical law may not be written in tensor form and nevertheless be covariant. Carmeli [1982] synthesized three possible statements for the Principle of Covariance, which are not exactly equivalents:

1st) all coordinate systems are equally good for expressing the Physical Laws, and all of them should be treated equivalently;

2nd) the Physical Laws must be expressed in tensor form in a four dimensional Riemannian spacetime;

3rd) each Physical Law must possess the same form in all coordinate systems.

In this Chapter we introduce tensors and we demonstrate that each tensor equation, if valid in a particular coordinate system, must be valid in any other one.

1.2. Vectors and Tensors

Let S and S' represented two coordinate systems, arbitrarily chosen, whose coordinates are represented respectively by x^i and x'^i, where i varies from 1 to N, in N dimensional spaces.

If we compare both systems of coordinates we will find between S and S' , relations of the type:

$$x'^i = F^i(x^j) \,, \qquad (1.2.1)$$

where | $i = 1, \dots, N$

| $j = 1, \dots, N$
| F^i = one to one function, continuous with continuous derivatives of any order.

Taking the differential of relation (1.2.1) we obtain by the rules of differential calculus:

$$dx'^i = \sum_{j=1}^{N} \frac{\partial x'^i}{\partial x^j} dx^j. \qquad (1.2.2)$$

From now on, we shall adhere to Einstein's sum convention, under which when one index is repeated twice in the same term of an equation, a sum over all values of that index is assumed. For example, we shall re-write (1.2.2) as:

$$dx'^i = \frac{\partial x'^i}{\partial x^j} dx^j. \qquad (1.2.3)$$

Repeated indices are called in English language as "dummies".

We now introduce the symbol δ^i_j , called Kronecker's delta. It is defined by the following relations:

$$\delta^i_j = 1 \quad \text{if} \quad i = j,$$

$$\delta^i_j = 0 \quad \text{if} \quad i \neq j.$$

A very useful mathematical relation is the following one:

$$\frac{\partial x^i}{\partial x^j} = \delta^i_j \quad . \qquad (1.2.4)$$

The above relation expresses that different coordinates are independent of one another. Applying a well known rule from Calculus, we can write from (1.2.4) that:

$$\frac{\partial x^i}{\partial x'^j} \frac{\partial x'^j}{\partial x^k} = \delta^i_k \quad . \qquad (1.2.5)$$

Contravariant Vectors

Relation (1.2.3) defines the transformation law for the components of the displacement vector (dx^i) between two infinitesimally nearby points. Each set of N quantities V^i which transform likewise, is called the set of components of a contravariant vector, so that the prototype vector V'^i in system S' is related to V^j in S by

$$V'^i = \frac{\partial x'^i}{\partial x^j} V^j \quad . \qquad (1.2.6)$$

Contravariant Tensors of 2ND Rank

Let T^{ij} be a set of N^2 quantities which transform according to the law

$$T'^{ij} = \frac{\partial x'^i}{\partial x^k}\frac{\partial x'^j}{\partial x^l}T^{kl} \quad . \qquad (1.2.7)$$

The above set T^{ij} is said to represent a contravariant tensor of 2nd rank. By the same token, higher order tensors could be defined. A closed analysis by the reader should suffice to check that by multiplying two contravariant vectors U^i and V^j , we obtain a 2nd rank contravariant tensor, and so on. However not always one could decompose a given tensor of superior order into the product of vectors. The reader may check also that by the given definitions, a contravariant vector is a tensor of first rank. The rank zero tensor is called an invariant; let ϕ and ϕ' represent an invariant in S and S' coordinate systems: obviously,

$$\phi = \phi' \quad . \qquad (1.2.8)$$

Covariant Vectors

Suppose that $\phi = \phi\ (\ x^i\ )$, is an invariant in each point of space, but a function of the coordinates x^i. From Calculus comes the relation

$$\frac{\partial \phi}{\partial x'^i} = \frac{\partial \phi}{\partial x^j}\frac{\partial x^j}{\partial x'^i} \quad . \qquad (1.2.9)$$

The reader should not have any trouble for identifying the "vector" $\frac{\partial \phi}{\partial x^j}$ as the gradient of ϕ . Each set of quantities V_i that transforms according to the gradient, i.e., according to (1.2.9), so that:

$$V'_j = \frac{\partial x^i}{\partial x'^j}V_i \quad , \qquad (1.2.10)$$

is called the set of components of a "covariant" vector, also called covector. As is clear in our notation, we adopt the convention of representing superior indices as a sign of contravariancy, while down indices represent covariance.

Covariant Tensors of 2nd Rank

A set of quantities T_{ij} is called the set of components of a 2nd rank covariant tensor if when transformed to the system S' we obtain the set T'_{ij} defined as:

$$T'_{ij} = \frac{\partial x^k}{\partial x'^i}\frac{\partial x^l}{\partial x'^j}T_{kl} \quad . \qquad (1.2.11)$$

Covariant tensors of higher order are defined analogously. The reader can check that the covariant tensor of 1st rank is a covector. The invariant is a covariant tensor of zero rank. In other words, the invariant is both covariant and contravariant. Like in the contravariant tensor case, the product of a covariant vector with components V_i with another covariant vector U_j yields a covariant tensor of 2nd rank but it is useful to remember that not all covariant tensors of 2nd rank can be decomposed into the product of two covariant vectors.

Mixed Tensors

Consider the product of a contravariant vector V^i with a covariant vector U_j:

$$T^i_j = V^i U_j \quad . \tag{1.2.12}$$

In another coordinate system T^i_j would be transformed into:

$$T'^i_j = \frac{\partial x'^i}{\partial x^k}\frac{\partial x^l}{x'^j} T^k_l \quad . \tag{1.2.13}$$

We say that the set T^i_j represents the components of a 2nd rank mixed tensor (one time covariant and other time contravariant).

In a similar way, mixed tensors of rank ($m+n$) can be defined, m times covariant and n times contravariant.

We invite the reader to show that Kronecker's delta is a mixed tensor of 2nd order. Hint: Notice the identity

$\delta^r_s V^s_j U^k \equiv V^r_j U^k$. The general rule is that when in a product with δ^r_s you can put $r = s$ and eliminate the delta.

Solved Problem

A covariant vector has components in the three-dimensional cartesian system of coordinates (x, y, z) given by:

$$\begin{aligned} T_x &= xyz \\ T_y &= y \\ T_z &= x \quad . \end{aligned} \tag{1.2.14}$$

It is asked to obtain the spherical components (T_r, T_θ, T_ϕ).

Solution:

In the notation of tensor calculus,

$$\begin{aligned} x^1 &= x \quad x'^1 = r \\ x^2 &= y \quad \text{and} \quad x'^2 = \theta \\ x^3 &= z \quad x'^3 = \phi \quad . \end{aligned} \tag{1.2.15}$$

According to the data given in the problem,

$$\begin{aligned} T_1 &= x^1 x^2 x^3 \\ T_2 &= x^2 \\ T_3 &= x^1 \ . \end{aligned} \tag{1.2.16}$$

The relation which express the cartesian coordinates as function of the spherical ones can be written as:

$$\begin{aligned} x^1 &= x'^1 senx'^2 . \cos x'^3 \\ x^2 &= x'^1 senx'^2 . senx'^3 \\ x^3 &= x'^1 \cos x'^2 \quad . \end{aligned} \tag{1.2.17}$$

We now recall the definition of a covariant vector, namely:

$$T'_j = \frac{\partial x^k}{\partial x'^j} T_k \quad .$$

$$\therefore$$

$$T_r \equiv T'_1 = \frac{\partial x^1}{\partial x'^1} T_1 + \frac{\partial x^2}{\partial x'^1} T_2 + \frac{\partial x^3}{\partial x'^1} T_3$$

$$T_\theta \equiv T'_2 = \frac{\partial x^1}{\partial x'^2} T_1 + \frac{\partial x^2}{\partial x'^2} T_2 + \frac{\partial x^3}{\partial x'^2} T_3$$
$$T_\phi \equiv T'_3 = \frac{\partial x^1}{\partial x'^3} T_1 + \frac{\partial x^2}{\partial x'^3} T_2 + \frac{\partial x^3}{\partial x'^3} T_3 .$$

On performing the calculations we obtain the solution of the problem. For instance:

$$\begin{aligned} T_r &= senx'^2 \cos x'^3 \; x^1 x^2 x^3 + senx'^2 senx'^3 \; x^2 + \cos x'^2 \; x^3 = \\ &= \text{sen } \theta \cos\phi \; (r \; sen\theta \cos\phi) \; (r \; sen\theta \; sen\phi) \; (r \cos\theta) + \\ &+ \text{sen } \theta \; sen\phi \; r \; sen\theta \; sen\phi + \cos\theta \; (r \cos\theta) = \\ &= r^3 (sen\theta)^3 \; (\cos\phi)^2 \; sen\phi \cos\theta + r \; (sen\theta)^2 \; (sen\phi)^2 + r \; (\cos\theta)^2 . \end{aligned}$$

1.3. Algebraic Operations with Tensors

Two tensors of the same rank and type may be added (or subtracted), yielding a new tensor of the same rank and type. The idea is that you only can sum bananas with bananas and apples with apples. By the type of the tensor we mean the precise number of contravariant indices and covariant indices. Example: given two 3rd rank tensors of the same type, say, A^i_{jk} and B^i_{jk} we obtain a new tensor

$$C^i_{jk} = A^i_{jk} + B^i_{jk} .$$

From the position of the indices it is clear that the sum is of the same type and rank. It has to be noticed once and for all that we speak of a tensor by referring to the set of its components which we represent by a typical member like C^i_{jk} that stands for the set.

FUNDAMENTAL THEOREM OF TENSOR CALCULUS:

Given a tensor equation valid in one system of coordinates, when we transform to another system of coordinates, both sides of the equation may change but we still get a valid equation.

Demonstration:

Let the given equation be valid in system S:

$$A^i_j = B^i_j \quad .$$

Transforming into the coordinate system S' it is obvious that both sides transform in the same way, *verbi gratia*, like a mixed tensor of 2nd rank (with other type of tensors the reasoning would not be any different), so that:

$$A'^i_j = B'^i_j \quad .$$

Simply stated, a tensor equation is valid in any system of coordinates. The mathematical reason behind this theorem is that tensor transformations are linear and homogeneous. Immediately one can conclude that any equation of the type $C^i_j = 0$ implies that

$C'^i_j = 0$, provided that we are dealing with a tensor.

(Remember that not all matrices of 2nd rank transform like tensors). In the above example if you write

$C^i_j = A^i_j - B^i_j = 0$, it is clear that $A'^i_j = B'^i_j$.

Exterior Product

The exterior product of two tensors was introduced earlier in the particular case when we commented that the product $V^j U_i$ of two vectors makes a 2nd rank tensor. A new tensor is always obtained which has a rank given by the sum of the ranks of the involved tensors in the exterior product, by the multiplication of the components of two or more tensors. As no summing is implicit in exterior product, care must be taken in order to represent each multiplying factor with different literal indices without repetition. For instance, suppose that we are given two tensors, say, A^r_{st} and B^{ik}_j. We multiply exteriorly to obtain a new tensor: C^{rik}_{stj} .

The value of each component of the new tensor will be given by:

$$C^{rik}_{stj} = A^r_{st} B^{ik}_j \quad .$$

Contraction

Let a given mixed tensor be contracted which means that we write one index of covariant type equal to a given contravariant index: we claim that the resultant tensor has a rank two units less than the original rank. In other words, the contracted tensor has a rank which results from computing the left untouched indices.

Demonstration: consider the tensor A^i_{jkl}. By definition the law of transformation is:

$$A'^i_{jkl} = \frac{\partial x'^i}{\partial x^r} \frac{\partial x^s}{\partial x'^j} \frac{\partial x^t}{\partial x'^k} \frac{\partial x^p}{\partial x'^l} A^r_{stp} \quad .$$

On contracting, i.e., writing $i = j$ we are left with:

$$A'^{i}_{ikl} = \left(\frac{\partial x'^i}{\partial x^r}\frac{\partial x^s}{\partial x'^i}\right)\frac{\partial x^t}{\partial x'^k}\frac{\partial x^p}{\partial x^l}A^r_{stp} = \frac{\partial x^t}{\partial x'^k}\frac{\partial x^p}{\partial x^l}A^r_{rtp} \quad , \quad \text{where we made use of (1.2.5).}$$

Q.E.D.

Interior Product

We define the interior product of two tensors by the exterior product followed by a contraction.

Tests of Tensor Character

Given a certain set of quantities, we would like to test the set in order to check whether we are in face of a tensor. The direct test is to check how the given set of quantities behaves under a coordinate transformation, say, from S to S'. Nevertheless, sometimes it is more useful to apply the so called Quotient Theorem which allows an indirect test: "if the interior product of the set "X" with an arbitrary tensor is a tensor, then X is a tensor". We leave it to the reader to demonstrate this theorem for a particular tensor like a 2nd rank contravariant. In other cases, the proof runs equally well.

1.4. Final Observations

1) **Transitivity**- Besides linearity and homogeneity the transformation property of tensors is endowed with transitivity. By transitivity we mean the following: if in a transformation from the coordinate system S to S' a given set X transforms according to the tensor definition, and the same happens from coordinate system S' to a third system S'' we are sure that the tensor character will be assured again. If this does not happen, the tensor definition would not be transitive. Let us demonstrate it for a 2nd rank contravariant tensor T^{ij} : the hypotheses are:

$$T'^{ij} = \frac{\partial x'^i}{\partial x^k}\frac{\partial x'^j}{\partial x^l}T^{kl} \quad . \qquad (1.4.1)$$

and,

$$T''^{s} = \frac{\partial x''^r}{\partial x'^i}\frac{\partial x''^s}{\partial x'^j}T'^{ij} \quad . \qquad (1.4.2)$$

The thesis is:

$$T'' = \frac{\partial x''^r}{\partial x^k}\frac{\partial x''^s}{\partial x^l}T^{kl} \quad . \qquad (1.4.3)$$

We demonstrate easily that substituting (1.4.2) into (1.4.1), and by applying twice the relation (1.2.5) we obtain (1.4.3).

2) **Tensorial Field**: The tensors defined in this Chapter refer to points in an amorphous space (Synge and Schild (1969)), what means that in this space there is no definition of distance between neighborhood points (the metric of the space is not necessarily defined

yet). Everything which is valid for isolated points in space, may be valid for a continuum, in which case we define a tensor field.

3) **Symmetry and Anti-symmetry**: A tensor of any order is said symmetric relative to a pair of indices of the same type, if the numerical value is not altered by an interchanged of those indices.

Example: if $T^{rsl}_{pq} = T^{srl}_{pq}$ then the tensor is symmetric relative to the first two contravariant indices.

We define an anti-symmetric tensor, relative to a pair of indices of the same type, if upon exchange of those indices, the components change the algebraic sign.

Example: suppose $T^{rsl}_{pq} = -T^{rsl}_{qp}$; then the tensor is anti-symmetric relative to the two covariant indices.

Anti-symmetric tensors are also called skew-symmetric tensors.

The reader may easily demonstrate that symmetry or anti-symmetry properties are conserved in a transformation of coordinates, for a tensor. This does not imply that an equality of the type:

$$T^{rsl}_{pq} = T^{psl}_{rq} \quad ,$$

will be kept under a transformation of coordinates. As an exercise the reader should show that the above is true.

In the next chapter use will be made of the following Theorem: "let $\Phi = g_{ij}X^iX^j$ be an invariant while X^i and X^j are arbitrary contravariant vectors and g_{ij} is a set of symmetric quantities in all system of coordinates. Then, g_{ij} stands for a 2nd rank covariant tensor".

4) **Riemannian Spaces:** For the reader versed in differential geometry we want to advance the idea that we shall need to particularize the otherwise amorphous space of this Chapter into a metric Riemannian space which is torsionless, where the affinities are the Christoffel symbols and there is a metric tensor of 2nd rank which is symmetric. The whole story is left for next Chapter.

References for Chapter 1

At this stage, the reader may benefit from consulting more advanced treatises like Weinberg (1972), Adler, Bazin and Schiffer (1975) or MTW (1973). These references are basic and highly recommended reading.

Chapter 2

Tensors in Riemann Spaces

2.1. Metric

In Euclidean Geometry, it is true that the distance between infinitesimally nearby points is an invariant. Choosing Cartesian orthogonal coordinates, we shall have:

$$ds^2 = dx^i dx^i \quad = \quad \text{invariant}, \qquad (2.1.1)$$

where a summation from 1 to 3 is implied by the repeated index i.

In curvilinear coordinates, ds^2 , called in general the metric form or the fundamental space form or the square of the line element, is given by an expression of the type:

$$ds^2 = g_{ij} dx^i dx^j \quad . \qquad (2.1.2)$$

For instance, in spherical coordinates, we shall have:

$$\begin{aligned} g_{11} &= 1, \\ g_{22} &= r^2 \ , \\ g_{33} &= r^2 sen^2\theta \ , \\ g_{ij} &= 0 \ , \text{for } i \neq j \ . \end{aligned} \qquad (2.1.3)$$

The 2nd rank covariant tensor, g_{ij} , is called covariant metric tensor or the fundamental space tensor. Notice that we may consider, *a priori*, the tensor g_{ij} to be symmetric, having in mind the commutativity of the product $dx^i dx^j$, and the subsequent verification that in expressions (2.1.2), only occur sums of the type

$$(g_{ij} + g_{ji}) \ .$$

Notice also that the proof that g_{ij} is a 2nd rank covariant tensor is obtainable from last Chapter theorem from Section I.4 where we take

$$\Phi = ds^2 \ ,$$

$$X^i = dx^i \ .$$

We say that the metric is indefinite whenever it may assume positive, negative or null numerical values. In the case of General Relativity Theory, as well as in the particular case where there is no gravitational field (Special Relativity) this can indeed happen. Recall Minkowski's spacetime metric:

$$ds^2 = -\left[dx^2 + dy^2 + dz^2\right] + c^2 dt^2, \qquad (2.1.4)$$

where c stands for the speed of light in vacuum for an inertial observer, and (x, y, z) are the space coordinates while t is the temporal coordinate. It is evident that the metric may be positive, negative or null. If $ds^2 = 0$, we say that the metric defines a null displacement, characteristic of light rays.

Riemann spaces are spaces that admit a metric of the type (2.1.2) where the covariant metric tensor is symmetric ($g_{ij} = g_{ji}$) .

Contravariant Metric Tensor

On considering g_{ij} like a square matrix, we can define a new tensor g^{ij} from the matrix above by the relation:

$$g^{ij} = \frac{\Delta^{ij}}{g} \ , \qquad (2.1.4)$$

where

Δ^{ij} = cofactor matrix for g_{ij}
$g =$ determinant of g_{ij} .

Taken that g_{ij} is symmetric, and from the definition of a matrix inverse, we find that:

$$g_{ij} g^{jk} = \delta_i^k \ . \qquad (2.1.5)$$

The tensor g^{jk} is sometimes called the conjugate tensor of g_{jk} .

In Riemannian spaces, distinct from amorphous spaces, there is no distinction among the covariant tensors, the contravariants or the mixed, when they refer to the same geometric entity. In fact, given any tensor T_k^{ij} we associate to it the tensor

$$S^{ijk} = g^{lk} T_l^{ij} \ . \qquad (2.1.6)$$

Now, let us associate to the tensor S^{ijk} a new tensor:

$$R_k^{ij} = g_{lk} S^{ijl} \ . \qquad (2.1.7)$$

A close examination reveals that

$$R_k^{ij} \equiv T_k^{ij} \quad . \qquad (2.1.8)$$

We conclude, therefore, that the metric tensors work as operators that lower down an index or raise up an index. When we apply this operator once there is no new tensor being created; however we have another representation of the same geometrical entity. In order to ensure proper mathematical consistency we shall use the same letter to represent the same entity with some raised or lowered indices, for example T_k^{ij} and T^{ijk} .

From relation (2.1.5) it is evident that Kronecker's delta is the mixed form of the metric tensor.

2.2. Christoffel Symbols

We define now Christoffel symbols of the 1st and 2nd kind, respectively:

$$\Gamma_{ijk} = \frac{1}{2}\left(\frac{\partial g_{ik}}{\partial x^j} + \frac{\partial g_{kj}}{\partial x^i} - \frac{\partial g_{ij}}{\partial x^k}\right) \quad , \qquad (2.2.1)$$

$$\Gamma^i_{jk} = g^{il}\Gamma_{jkl} \quad . \qquad (2.2.2)$$

Notwithstanding the fact that the symbols Γ_{ijk} and Γ^i_{jk} contain indices, they do not represent tensors as we shall show below.

We shall now show that in a change of coordinates from system S to system S' the Christoffel symbol of the 1st kind Γ_{ijk} transforms itself into Γ'_{ijk} given by the following expression:

$$\Gamma'_{ijk} = \Gamma_{lmn}\frac{\partial x^l}{\partial x'^i}\frac{\partial x^m}{\partial x'^j}\frac{\partial x^n}{\partial x'^k} + g_{lm}\frac{\partial x^l}{\partial x'^k}\frac{\partial^2 x^m}{\partial x'^i \partial x'^j} \quad . \qquad (2.2.3)$$

Before we derive this relation, we must point out that it is not in the form of a tensor transformation: the right hand side of the above formula contains two terms summed, the 1st of which if stood alone, would warrant the tensorial nature of the Christoffel symbol of the 1st kind and consequently, of the 2nd kind too.

Demonstration:

From the definition property of a covariant 2nd rank tensor applied to the metric tensor:

$$g'_{jk} = \frac{\partial x^l}{\partial x'^j}\frac{\partial x^m}{\partial x'^k} g_{lm} \quad . \qquad (2.2.4)$$

Applying the partial derivative symbol relative to x'^n to both sides of the above equation, we find by the usual Differential Calculus rules:

$$\frac{\partial g'_{jk}}{\partial x'^n} = \frac{\partial x^l}{\partial x'^j}\frac{\partial x^m}{\partial x'^k}\frac{\partial x^p}{\partial x'^n}\frac{\partial g_{lm}}{\partial x^p} + \frac{\partial^2 x^l}{\partial x'^n \partial x'^j}\frac{\partial x^m}{\partial x'^k} g_{lm} + \frac{\partial^2 x^m}{\partial x'^n \partial x'^k} g_{lm} \frac{\partial x^l}{\partial x'^j} \quad . \tag{2.2.5}$$

By applying repeatedly relation (2.2.5) to the definition of Γ'_{ijk} , i.e., the same definition in the coordinate system S' , and remembering the definition of the 1st kind symbols in the system S we obtain result (2.2.3) straightforwardly.

We could equally well obtain the transformation law for the gamma symbols of the 2nd kind and find:

$$\Gamma'^i_{jk} = \Gamma^l_{mn} \frac{\partial x'^i}{\partial x^l}\frac{\partial x^m}{\partial x'^j}\frac{\partial x^n}{\partial x'^k} + \frac{\partial x'^i}{\partial x^n}\frac{\partial^2 x^n}{\partial x'^j \partial x'^k} \quad . \tag{2.2.6}$$

We remark that in the above expression, the first rhs term is the one that, on standing alone would guarantee the tensor character of the gammas, which is not the case here because of the 2nd term in the rhs.

2.3. Covariant Derivative

Except for the case of the derivative of an invariant, in which case we obtain a gradient (a covariant vector) , all other partial derivatives of tensors do not result in other tensors. Let us show an example of the above with a contravariant vector V^i . By definition: $V'^i = \frac{\partial x'^i}{\partial x^j} V^j$.

On taking partial derivatives, say relative to x'^k with find:

$$\frac{\partial V'^i}{\partial x'^k} = \frac{\partial x'^i}{\partial x^j}\frac{\partial V^j}{\partial x'^k} + \frac{\partial^2 x'^i}{\partial x'^k \partial x^j} V^j = \frac{\partial x'^i}{\partial x^j}\frac{\partial x^l}{\partial x'^k}\frac{\partial V^j}{\partial x^l} + \frac{\partial^2 x'^i}{\partial x^l \partial x^j}\frac{\partial x^l}{\partial x'^k} V^j \quad . \tag{2.3.1}$$

The first term of the right hand side of (2.3.1) should be the unique to exist if the quantity in the left hand side was a tensor, in this case a second rank mixed tensor. The existence of second derivatives can nevertheless be used to define another kind of derivative, called "covariant derivative" that we shall represent by $\frac{DV^i}{Dx^k}$: it will be equal to the simple partial derivative $\frac{\partial V^i}{\partial x^k}$ plus a corrective term such that the sum will be a tensor.

In order to get the said definition, we first notice that from the law transformation of the Christoffel symbols (2.2.6) the following relation is valid:

$$\frac{\partial^2 x'^i}{\partial x^j \partial x^j} = \Gamma^k_{jl} \frac{\partial x'^i}{\partial x^k} - \frac{\partial x'^m}{\partial x^j}\frac{\partial x'^k}{\partial x^l}\Gamma'^i_{mk} \quad . \tag{2.3.2}$$

We now plug the above result into (2.2.7) and we obtain:

$$\left[\frac{\partial V'^i}{\partial x'^k} + \Gamma'^i_{kl} V'^l\right] = \frac{\partial x'^i}{\partial x^j}\frac{\partial x^l}{\partial x'^k}\left[\frac{\partial V^j}{\partial x^l} + \Gamma^j_{ls} V^s\right] \quad . \tag{2.3.3}$$

We define thus, the covariant derivative of the vector V^j relative to x^l, as:

$$\frac{DV^j}{Dx^l} \equiv \frac{\partial V^j}{\partial x^l} + \Gamma^j_{ls} V^s \, . \tag{2.3.4}$$

Other usual notations are:

$$\frac{DV^j}{Dx^l} \equiv \nabla_l V^j \equiv V^j_{;l} \quad ,$$

while the partial derivatives appear with a comma, like

$$\frac{\partial V^j}{\partial x^l} \equiv V^j_{,l} \ .$$

By means of (2.3.2), we can immediately obtain a definition for the covariant derivative of a covariant vector:

$$V_{j;k} = \frac{\partial V_j}{\partial x^k} - \Gamma^l_{jk} V_l \quad . \qquad (2.3.5)$$

In order to obtain covariant derivatives for 2nd rank tensors it suffices to imagine how would be the covariant derivative of the product of two vectors, equal to the given tensor: then we apply the formulae (2.3.4) and (2.3.5) obtaining straightforwardly the result:

$$T^{ij}_{;k} = \frac{\partial T^{ij}}{\partial x^k} + \Gamma^i_{kl} T^{lj} + \Gamma^j_{kl} T^{il} \qquad (2.3.6)$$

$$T_{ij;k} = \frac{\partial T_{ij}}{\partial x^k} - \Gamma^l_{ki} T_{lj} - \Gamma^l_{kj} T_{il} \qquad (2.3.7)$$

$$T^i_{j;k} = \frac{\partial T^i_j}{\partial x^k} + \Gamma^i_{kl} T^l_j - \Gamma^l_{kj} T^i_l \ . \qquad (2.3.8)$$

Higher order covariant derivatives and covariant derivatives of higher rank tensors, can be obtained by extending the above techniques. We can easily infer general rules for those cases which are left to the reader. We also leave up to the reader to show that the usual rules for derivatives of sums and products are the same in the case of covariant derivatives.

We now show one of the most important results of Riemannian Geometry , *verbi gratia*, that the covariant derivative of the metric tensor is always null. We ask the reader to combine relations (2.3.6), (2.3.7), (2.2.1) and (2.2.2) in order to prove that:

$$g_{ij;k} = 0 \text{ , and,} \qquad (2.3.9)$$

$$g^{ij}_{;k} = 0 \quad . \qquad (2.3.10)$$

2.4. Curvature Tensors

For simple derivatives, the following commutation relation is valid:

$$\frac{\partial}{\partial x^\alpha}\frac{\partial}{\partial x^\beta} = \frac{\partial}{\partial x^\beta}\frac{\partial}{\partial x^\alpha} \ .$$

We should not expect however that the covariant derivative can commute in the same way as above in the general case of a Riemannian space:

$$V^i_{;j;k} \neq V^i_{;k;j} \quad .$$

Its very easy to show that:

$$V^i_{;j;k} - V^i_{;k;j} = -R^i_{lkj} V^l \qquad (2.4.1)$$

and

$$V_{i;j;k} - V_{i;k;j} = R^l_{ijk} V_l \qquad (2.4.2)$$

where

$$R^l_{ijk} \equiv \frac{\partial \Gamma^l_{ik}}{\partial x^j} - \frac{\partial \Gamma^l_{ji}}{\partial x^k} + \Gamma^m_{ik} \Gamma^l_{jm} - \Gamma^m_{ij} \Gamma^l_{km} \, . \qquad (2.4.3)$$

The proof of the above is direct: one has only to apply relations (2.3.6) and (2.3.7), combined with (2.3.4) and (2.3.5).

The fourth rank tensor R^i_{jkl} is the famous curvature tensor or Riemann-Christoffel tensor.

Ricci Tensor

By definition, the Ricci Tensor R_{ij} is given by:

$$R_{ij} = R^l_{ilj} \quad . \qquad (2.4.4)$$

Ricci Scalar Curvature

By definition, the Ricci Scalar Curvature R is given by:

$$R \equiv R^i_i \equiv g^{ij} R_{ij} \quad . \qquad (2.4.5)$$

Einstein's Tensor

By definition, Einstein's Tensor G_{ij} is given by:

$$G_{ij} = R_{ij} - \frac{1}{2} g_{ij} R \quad . \qquad (2.4.6)$$

Geodesic Coordinate System

A coordinate system is called geodesic, if in a considered point of spacetime, all the components of the Christoffel symbols are null.

Theorem:

"It is always possible, in a given point in spacetime, to build a geodesic coordinate system".

We now demonstrate this theorem:

Consider a coordinate system S in which, for a given spacetime point the gammas are different from zero: $\Gamma^i_{jk} \neq 0$.

We now go to another system of coordinates S' which has the following property:

$$x'^i = x^i - x^i_0 + \tfrac{1}{2}\left(\Gamma^i_{jk}\right)_0 \left[x^j - x^j_0\right]\left[x^k - x^k_0\right] \quad . \tag{2.4.7}$$

The index zero indicates the values of the quantities in the point where the geodesic coordinate system is defined.

We now obtain the null $\left(\Gamma'_{ijk}\right)_0$ by means of relation (2.2.6) if we introduce the following results obtained from (2.4.7):

$$\left(\frac{\partial x'^i}{\partial x^j}\right)_0 = \delta^i_j$$

$$\left(\frac{\partial x^i}{\partial x'^j}\right)_0 = \delta^i_j$$

$$\left(\frac{\partial^2 x^i}{\partial x'^j \partial x'^k}\right)_0 = \left(-\Gamma^i_{jk}\right)_0 \quad .$$

Q.E.D.

Theorem:
"In a geodesic system, the partial derivatives of the metric tensor are all null".

Demonstration

We know that the covariant derivatives of the metric tensor are null; however in a geodesic system of coordinates the covariant derivative of the metric tensor equals the partial derivative, so these are also null.

Bianchi Identity

In a geodesic system of coordinates, all the Christoffel symbols are null, but not necessarily the derivatives are null too. In such case we may write:

$$R^i_{jkl;m} = \frac{\partial^2 \Gamma^i_{jl}}{\partial x^m \partial x^k} - \frac{\partial^2 \Gamma^i_{jk}}{\partial x^m \partial x^l} \quad . \tag{2.4.8}$$

On applying the above equation three times, we obtain:

$$R^i_{jkl;m} + R^i_{jlm;k} + R^i_{jmk;l} = 0 \quad . \tag{2.4.9}$$

This is a tensorial equation valid for a geodesic coordinate system. However, according to the Fundamental Theorem of Tensor Calculus, derived in Chapter 1, it must be valid in any system: it is called Bianchi Identity.

We now go ahead to perform the most important calculation involving Einstein's tensor: we show that the covariant divergence of the Einstein Tensor is null:

$$G^{i}_{j;i} = 0 \quad .$$

Before doing the calculation, one word of caution is necessary: when speaking about the divergence of a tensor, in Riemannian space, the covariant divergence is implied always. Recall now Bianchi identity (2.4.8), which we now cast in the form:

$$R_{pjkl;m} + R_{pjmk;l} + R_{pjlm;k} = 0 \quad . \tag{2.4.10}$$

The last equation is obtained from (2.4.8) by multiplying both sides of the equation by g_{pi} and remembering that the covariant derivative of the metric tensor is zero. On the same token, we now multiply by g^{pk} and obtain, given the definition of the Ricci tensor (2.4.4) that:

$$R_{jl;m} - R_{jm;l} + R^{k}_{jlm;k} = 0.$$

We now contract indices by multiplication with g^{jl} and get:

$$R_{;m} - R^{l}_{m;l} - R^{k}_{m;k} = 0.$$

The last two terms are equal because the indices "l" and "k" are dummies. We then find:

$$\left(R^{k}_{m} - \tfrac{1}{2}\delta^{k}_{m}R\right)_{;k} = 0.$$

Q.E.D

Now we shall make our first encounter with the cosmological constant (lambda): on remembering that the covariant derivative of a scalar like R , is equal to the usual derivative, we may write with impunity the following relation which is equivalent to the above:

$$\left[R^{k}_{m} - \tfrac{1}{2}\delta^{k}_{m}\left(R - 2\Lambda\right)\right]_{;k} = 0 \quad . \tag{2.4.11}$$

Albert Einstein was led to introduce the lambda constant into the Theory of General Relativity as a mathematical artifact which allowed him to obtain, with a proper adjustment of the value of lambda, a cosmological solution for the Universe that turned it static. We shall further study the consequences of the introduction of the "cosmological" constant in later Chapters.

References for Further Study

Same as in last Chapter. See also Raychaudhuri et al (1992) and Ray d'Inverno (1993).

Part II

Introduction to General Relativity

Chapter 3

Basic Theory

3.1. The Basic Principles

Should a theory be "beautiful" in order to be acceptable? — In my subjective opinion, YES! For women, intelligence does the job. South American judges and politicians are usually clever; they get rich, but they are seldom intelligent. Some are even crazy. Also, some scientists (but not me !). Let the ugly ones, forgive me, because for theories, beauty is fundamental.

General Relativity Theory is an intelligent theory of Gravitation, endowed with beautiful equations. It acknowledges that, in a given point of space, an accelerated system is equivalent to a local inertial one, subject to a gravitational field (Principle of Equivalence). This principle has experimental basis in Eötvös experiment, which suggests the equality between inertial and gravitational mass. In Newtonian Mechanics the origin of the inertial mass comes from the second Newton's Law (forces are proportional to accelerations; the constant of proportionality is inertial mass); gravitational mass, is the one which makes the weight of a body proportional to the acceleration of gravity (the constant of proportionality being the gravitational mass).

GRT obeys the covariance principle for the laws of Physics, as already commented. The father of the theory, Albert Einstein, tried to find equations for the motion of bodies in the presence of gravitational fields, which he expected to reduce to the known laws of Special Relativity, when the gravitational field would be switched off; and in the case of weak gravitational fields, and low speeds, when compared to the speed of light, Einstein expected that the GRT equations reduced to those of Newtonian gravitation.

We shall see later that a particle's acceleration in a gravitational field, is proportional to the Christoffel symbols. As it was showed in last chapter, it is always possible, by means of a coordinate transformation (to a geodesic system), to turn to zero the Christoffel symbols at a given point of spacetime. Then, in the new localized coordinate system, at that point, a body feels no acceleration: this makes sense with the Equivalence Principle. For more on "principles", we refer to Chapter 6.

3.2. Energy-Momentum Tensor

For a continuum medium, it can be shown within the Special Relativity Theory, which is the valid theory in the absence of gravitation, that there exists a symmetric tensor, T'^{ij} , so that:

$$\frac{\partial T'^{ij}}{\partial x^i} = D^j \quad , \qquad (3.2.1)$$

where D^j stands for the external quadriforce acting on the considered system. If the system is free from external forces, – or if it is not, and then, we include the external environment into the old system, so that the prior external forces become now internal, – we know that we can find a new tensor T^{ij} called the energy momentum tensor of the system, so that its usual divergence is null:

$$\frac{\partial T^{ij}}{\partial x^i} = 0 \quad . \qquad (3.2.2)$$

The above equation works for a isolated system, and we shall show below that, for a perfect fluid, the (3.2.2) equations reduce themselves to the known continuity and Euler equations of Newtonian Mechanics. For a perfect fluid, we find tentatively an expression for the energy tensor that obeys what we said above. We now offer the final result,

$$T^{ij} = \left[\left(\rho + \frac{p}{c^2}\right) u^i u^j - g^{ij} \frac{p}{c^2}\right] \quad , \qquad (3.2.3)$$

where ρ is the energy density of the fluid. p is the pressure and u^i the quadrivelocity of the fluid, defined by

$$u^i = \frac{dx^i}{ds} \quad ,$$

where c is the speed of light in vacuum for an inertial observer, and the ds^2 is the metric of Special Relativity:

$$ds^2 = -dx^\nu dx^\nu + \left(dx^0\right)^2 = -dx^\nu dx^\nu + c^2 dt^2 \quad . \qquad (3.2.4)$$

Notice that Greek indices are used here, for three dimensional spaces; for instance ν =1,2,3. When we reduce to Newtonian Gravitation we must obey the following limits:

$$u^\nu \longrightarrow \frac{v^\nu}{c} = \frac{dx^\nu}{cdt}$$

$$u^0 \longrightarrow 1 \qquad (3.2.5)$$

$$c \longrightarrow \infty$$

$$t \equiv t' \quad .$$

For more details, consult a Special Relativity book like Synge (1965). Now we show that our choice for T^{ij} as given by (3.2.3) for a perfect fluid, reduce to the well known continuity and Euler equations.

1st) Continuity Equation.

From (3.2.2), given the symmetry of T^{ij} we obtain:

$$\frac{\partial T^{0i}}{\partial x^i} = 0 \quad . \tag{3.2.6}$$

From the above equation we obtain:

$$\frac{\partial}{\partial x^i}\left[\left(\rho + \frac{p}{c^2}\right) u^0 u^i - g^{0i}\frac{p}{c^2}\right] = 0 \ .$$

From (3.2.4) we find $g^{0i} = g^{00} = 1$ if $i = 0$.

We also find that $g^{0i} = 0$ if $i \neq 0$.

Therefore we shall find

$$\frac{\partial}{\partial x^\nu}\left[\left(\rho + \frac{p}{c^2}\right) u^0 u^\nu - \frac{p}{c^2}\right] + \frac{\partial}{\partial x^0}\left[\left(\rho + \frac{p}{c^2} - \frac{p}{c^2}\right)\right] = 0 \, .$$

In the limit (3.2.5) we obtain:

$$\frac{\partial}{\partial x^\nu}\left(\rho \frac{v^\nu}{c}\right) + \frac{1}{c}\frac{\partial \rho}{\partial t} = 0$$

which can also be simplified into:

$$\frac{\partial}{\partial x^\nu}\left(\rho v^\nu\right) + \frac{\partial \rho}{\partial t} = 0 \ ,$$

which is the famous continuity equation of fluid dynamics in the Newtonian case.

2nd) Euler's equation.

We obtained (3.2.6) by making $j = 0$ in (3.2.2). Likewise, we now make

$$j = i = 0, 1, 2, 3 \ , \quad \text{or} \quad \frac{\partial T^{i\nu}}{\partial x^i} = 0 \quad . \tag{3.2.7}$$

We find explicitly:

$$\frac{\partial}{\partial x^\nu}\left[\left(\rho + \frac{p}{c^2}\right) u^\nu u^\mu - g^{\mu\nu}\frac{p}{c^2}\right] + \frac{\partial}{\partial x^0}\left[\left(\rho + \frac{p}{c^2}\right) u^\nu u^0 - g^{\mu 0}\frac{p}{c^2}\right] = 0.$$

Going to the desired limit:

$$\frac{\partial}{\partial x^\nu}\left[\rho \frac{v^\nu v^\mu}{c^2} - g^{\mu\nu}\frac{p}{c^2}\right] + \frac{1}{c}\frac{\partial}{\partial t}\left(\rho \frac{v^\mu}{c^2}\right) = 0.$$

We now multiply all the terms by c^2 , and taking to account that $g^{\mu\nu}$ reduces to the delta of Kronecker - $\delta^{\mu\nu}$ we obtain:

$$\frac{\partial}{\partial t}\left(\rho v^\mu\right) + \frac{\partial}{\partial x^\nu}\left[\rho v^\nu v^\nu + \delta^{\mu\nu} p\right] = 0 \quad . \tag{3.2.8}$$

This is Euler's equation which is essentially the fluid dynamical second Newton's Law as we shall now show.

Newton's second Law for perfect fluids can be written:

$$\frac{d}{dt}(\rho\vec{v}) = -grad\ p = -\nabla p \quad . \tag{3.2.9}$$

From the Differential Calculus of many variables we find that:

$$\frac{d}{dt}(\rho\vec{v}) \equiv \frac{\partial}{\partial t}(\rho\vec{v}) + \frac{\partial}{\partial x^\nu}(\rho\vec{v})\, v^\nu \quad . \tag{3.2.10}$$

Now

$$v^\nu = \frac{dx^\nu}{dt}$$

(This is Newtonian definition of velocity).

We now rewrite (3.2.9) by means of (3.2.10) in a particular generic component v^μ of velocity and we retrieve equation (3.2.8) QED.

3.3. T^{ij} for the Electromagnetic Field

Maxwell's equations for the electromagnetic field are given by

$$\begin{aligned} \text{div}\ \vec{E} &= \rho_e \ , \\ \text{rot}\ \vec{B} &= \frac{\partial\vec{E}}{\partial t} + \vec{J} \ , \\ \text{div}\ \vec{B} &= 0 \ , \\ \text{rot}\ \vec{E} &= -\frac{\partial\vec{B}}{\partial t} \ , \end{aligned} \tag{3.3.1}$$

where $\vec{E}$ and $\vec{B}$ stands for the electric and magnetic fields, $\vec{J}$ and ρ_e are the densities of electric current and charge. College students who find strange the above equations may benefit from Purcell (1985), Chapter 9.

We define now the Electromagnetic field tensor F^{ij} by giving its components in terms of the above fields:

$$\begin{aligned} F^{12} &= B_3 \\ F^{01} &= E_1 \\ F^{23} &= B_1 \\ F^{02} &= E_2 \\ F^{31} &= B_2 \end{aligned}$$

$$F^{03} = E_3$$

$$F^{ij} = -F^{ji} .$$

The reader may easily demonstrate that Maxwell's equations in terms of the antisymmetric tensor F^{ij} become:

$$\frac{\partial}{\partial x^i} F^{ij} = -J^j , \qquad (3.3.2a)$$

and

$$\frac{\partial}{\partial x^i} F_{jk} + \frac{\partial}{\partial x^j} F_{ki} + \frac{\partial}{\partial x^k} F_{ij} = 0 , \qquad (3.3.2b)$$

where

$$J^0 = \rho_e .$$

Textbooks in Special Relativity usually demonstrate that, for the electromagnetic field, the pure energy-momentum tensor, as we defined earlier, is endowed with property (3.2.2), namely, that the divergence of the energy momentum tensor is null, and that it is given by:

$$T^{ij} = F^i_k F^{jk} + \frac{1}{4} g^{ij} F_{kl} F^{kl} . \qquad (3.3.3)$$

For details on this and other matters, consult Synge (1965) or references in Chapter 1.

A special comment must be made with reference to the electromagnetic energy tensor field: we defined it in terms of its components in the reference system where Maxwell's equations were cast; it is an experimental result, that the physical quantities represented by it, change from one coordinate system into another, in such a way that electromagnetic energy tensor field is really deserving the word "tensor", .i.e. for example, relations (3.3.3) and (3.2.2), are true in any coordinate system.

3.4. Einstein's Field Equations

Before going to Einstein's field equations, we must remember that the obvious generalization of property (3.2.2) which is valid for Minkowski's spacetime, would be to introduce covariant derivatives instead of simple derivatives when gravitation is present and Riemannian geometry takes place. So reasoned Albert Einstein, and the key idea was that the field equations of General Relativity should reduce to the equations of Special Relativity in the absence of gravitation. All equations should be written in tensor form and for weak gravitational fields they should reduce to Newtonian gravity laws. So, in the presence of gravitational fields, the covariant divergence of the energy momentum tensor, should be conserved, in the covariant derivative sense:

$$T^{ij}_{;j} = 0. \qquad (3.4.1)$$

If the equation for matter fields, is written with the tensor T^{ij} obeying the above property and on remembering that in Riemannian geometry which has a similar law of conservation for the Einstein's tensor, namely

$$G^{ij}_{;j} = 0, \qquad (3.4.2)$$

which we studied in Chapter 2, it would be reasonable that matter and geometry of space were tied by the equations

$$\boxed{G^{ij} = -\kappa T^{ij}} \qquad \text{(Einstein's field Equations)} \qquad (3.4.3)$$

because then the properties (3.4.2) and (3.4.1) would be automatically fulfilled: we may say that matter makes the geometry. Precisely speaking, if we know the energy density and the pressure of a fluid, we may build its energy momentum tensor, and then we shall know how this matter curves spacetime through the Einstein tensor, which reflects the properties of the curvature of spacetime. The constant κ is called Einstein's gravitational constant and it is closely related to Newtonian gravitational constant G as we shall show later on. The plausibility for (3.4.3), lies in the paragraph following formula (3.4.5) below.

Einstein's field equations should reduce for weak gravitational fields and low speeds when compared with the speed of light, to the Newtonian equations of Poisson:

$$\frac{\partial^2 \Psi}{\partial x^\nu \partial x^\nu} = -4\pi G \rho \qquad (3.4.4)$$

where Ψ is the Newtonian scalar potential, G is the Newton's gravitational constant and ρ stands for mass density, while the potential Ψ relates with Newton's second law by

$$\frac{d^2 x^\nu}{dt^2} = -\frac{\partial \Psi}{\partial x^\nu} \quad . \qquad (3.4.5)$$

In order that this last expression is also verified we need an additional hypothesis: a particle subject to a gravitational field moves according to a timelike geodesic of Riemann's spacetime. Roughly speaking, we can say that we have found the following reduction, which yields (3.4.4) from (3.4.3):

$$\boxed{G^{ij} \longrightarrow \frac{\partial^2 \Psi}{\partial x^\nu \partial x^\nu}} \quad \text{while} \quad \boxed{T^{ij} \longrightarrow -\tfrac{1}{2}\kappa\rho} \quad .$$

In Chapter 5, section 5.2, we shall see that the equation of geodesics expresses the fact that the tangent vector to the curve stays parallel to itself during motion.

The equation derived for geodesics in Riemann spaces is

$$\frac{d^2 x^i}{ds^2} + \Gamma^i_{jk} u^j u^k = 0 \quad . \qquad (3.4.6)$$

Now, we shall show that geodesics reduce to straight lines of Euclidean geometry in the absence of gravitation. Indeed, if gravitation is absent we can choose Cartesian coordinates where the Christoffel symbols are null. We are left with the equation:

$$\frac{d^2x^i}{ds^2} = 0.$$

For $i = 0$ we find

$$\frac{d^2x^0}{ds^2} = 0$$

which has the solution $x^0 = s$. With this solution we find:

$$\frac{d^2x^i}{dx_0^2} = 0 \quad \text{or}$$

$$\frac{d^2x^i}{dt^2} = 0$$

which are the equations of a straight line. QED.

We remember that straight lines are the shortest curves between two points in Euclidean Geometry and that a similar rôle can be attached to geodesics in the Riemannian geometry (stationary property).

Einstein proposed, in the year 1916, that the geodesic postulate should be incorporated to GRT as a separate one. Eleven years later however, Einstein and Grommer showed that this postulate could be indeed demonstrated with help of the field equations alone. Nevertheless, there are some restrictions for the derivation of this postulate from the field equations; for instance, a spinning particle does not follow geodesics. (see details in Papapetrou, 1974 – Section 42).

In this Chapter, we shall offer a particular case ($p = 0$ matter , also called "dust"), where we shall show how to derive the equation of geodesics from field equations (3.4.3).

3.5. Introducing the Cosmological Constant

The Einstein tensor, G^{ij}, as we recall from Chapter 2, had the property that its covariant divergence was null. Because any tensor proportional to the metric tensor has zero covariant derivative we may say that any tensor of the type:

$$E^{ij} = G^{ij} + \Lambda g^{ij} \quad , \qquad (3.5.1)$$

where Λ is an arbitrary constant, has also a null covariant divergence:

$$E^{ij}_{;j} = 0. \qquad (3.5.2)$$

Albert Einstein furthermore concluded that his field equations should be conceived with the lambda constant, employing thus the tensor E^{ij} instead of G^{ij} so that we would find:

$$\boxed{E^{ij} = -\kappa T^{ij}} \qquad (3.5.3)$$

We call the above field equations (3.5.3) as the field equations with a cosmological constant. The word "cosmological" is not without reason: for local Physics experiments we do not need such constant. It is only when large portions of the Universe are considered, that lambda is really needed. However, even in other fields, lambda may be important, as the energy density of the vacuum, as stated below.

We leave as an exercise to the reader to follow all the steps in order to demonstrate that, alternatively, for a perfect fluid energy-momentum tensor T^{ij} as given by (3.2.3), we can define a new energy-momentum tensor

$$\widetilde{T}^{ij} = \left(\widetilde{\rho} + \frac{\widetilde{p}}{c^2}\right) u^i u^j - g^{ij} \frac{\widetilde{p}}{c^2} \qquad (3.5.4)$$

where there is a redefinition of energy density, and fluid pressure, of the type:

$$\frac{\widetilde{p}}{c^2} = \frac{p}{c^2} - \frac{\Lambda}{\kappa} \quad ; \qquad (3.5.5)$$
$$\widetilde{\rho} = \rho + \frac{\Lambda}{\kappa}.$$

With this new energy-momentum tensor we can write:

$$G^{ij} = -\kappa \widetilde{T}^{ij} \quad . \qquad (3.5.6)$$

If the student finds this boring or confusing, be aware that we are dealing with a great problem in GRT: the cosmological constant problem. With the field equations (3.5.6) we have transferred the responsibility for keeping the account of the cosmological term from the Einstein's tensor E^{ij} to a modified energy-momentum tensor $\widetilde{T}^{ij}$ in such a way that we may imagine to have augmented the energy density by a parcel $\frac{\Lambda}{\kappa}$: the question remains, where does this energy come from? The Russian cosmologist Yakov Zel'dovich, in the year 1974, prompted an answer to this question: the additional energy comes - yes!!! - from the energy density of the vacuum. So, the zero field energy of Quantum Mechanics can find its way into General Relativity, which is a Classical Theory, through the lambda constant. Zel'dovich and others have hinted that besides the normal fluid for matter, the Universe is permeated with a second fluid which is offered to us "free of charge" and contributes with an energy density given by $\frac{\Lambda}{\kappa}$ and a negative pressure, $-\frac{\Lambda}{\kappa}$ as can be checked from the first relation numbered (3.5.5). Physicists today joke that Zel'dovich cleaned the Geometry from an annoying lambda and threw away the burden to the Physics part of the field equations, in the r.h.s.

New cosmological discoveries in the beginning of the twenty-first century showed that the Universe has been accelerating its expansion, which points to a tiny but non zero cosmological constant to pervade the equations of cosmology. Albert Einstein was led in some period of his life to regret having introduced the lambda constant, describing this fact as the greatest blunder in his life. Despite the fact that he never wrote such a thing, his contemporary colleague George Gamow recalled in his memories having heard Einstein complain in that way, about the lambda constant. It is refreshing to mention that Particle Physicists require that, in the very early Universe, the energy of the vacuum be extraordinarily huge. Nevertheless, we shall see in next Chapter that the Schwarzschild's solution for the solar system, while neglecting the cosmological term, is a very accurate description of astronomical observations near the Sun and the Earth, thus pointing to a very small lambda. In fact, it should be 10^{-120} orders of magnitude lower than what Particle Physics needs for the very early Universe. We shall see in further chapters, that some physicists, like Orfeu Bertolami and Marcelo Berman worked with the possibility of a lambda term not constant but decaying with the square of the age of the Universe. Alternatively Chen and Wu and Berman also worked with a solution for the lambda problem, that would make for a lambda term decaying with the square of the scale-factor (sometimes called the "radius") of the Universe. For other possibilities, see Overduin and Cooperstock (1998).

3.6. Reduction of GRT into Newtonian Physics

Our purpose in what follows shall be to prove that the Newtonian potential Ψ is related to the metric tensor, so that g^{ij} are representative of the gravitational potentials in Einstein's theory. Furthermore, it will be shown that the condition for the field equations (3.4.3) to reduce to the Newtonian Poisson equation is that:

$$\kappa = \frac{8\pi G}{c^2} \quad . \qquad (3.6.1)$$

Note that the field equations can be written with another sign in the right hand side, if κ is defined as $(-\kappa)$.

1st **Condition: Relations between g^{ij} and Ψ .**

Equation (3.4.6), the geodesic equations, do reduce in the static case and for weak gravitational fields, like it follows. A weak field differs very little from Minkowskian metric. Let us write

$$g_{ij} = \eta_{ij} + \gamma_{ij} \quad , \qquad (3.6.2)$$

where η_{ij} is a the diagonal matrix ($+1, -1, -1, -1$) and the γ_{ij}'s are very less than unity in absolute value. If we now impose a static metric, i.e.,

$$\frac{\partial g_{ij}}{\partial x^0} \approxeq 0 \quad ,$$

and if we consider low speeds in comparison with the speed of light, the (3.4.6) should reduce to

$$\frac{d^2x^i}{ds^2} + \Gamma^i_{00} \approxeq 0 \quad ,$$

where

$$\Gamma^i_{00} = \frac{1}{2} g^{ij} \left(\frac{\partial g_{0j}}{\partial x^0} + \frac{\partial g_{0j}}{\partial x^0} - \frac{\partial g_{00}}{\partial x^j} \right) \approxeq -\frac{1}{2} g^{ii} \frac{\partial g_{00}}{\partial x^i} \approxeq -\frac{1}{2} \frac{\partial \gamma_{00}}{\partial x^i} \approxeq -\frac{1}{2} \frac{\partial g_{00}}{\partial x^i} \ .$$

Then

$$\frac{d^2x^i}{dt^2} \approxeq -\frac{c^2}{2} \frac{\partial g_{00}}{\partial x^i} \ . \qquad (3.6.3)$$

On comparing (3.6.3) with the Newtonian equation:

$$\frac{d^2x^i}{dt^2} = -\frac{\partial \Psi}{\partial x^i} \quad ,$$

we find that:

$$\Psi \approxeq \frac{g_{00}c^2}{2} + \text{constant} .$$

In order to determine the value of the constant above we shall consider Minkowski's metric, which is the limiting metric for points that are infinitely far from the masses. In this case:

$$g_{00} \rightarrow 1 \quad ,$$

$$\Psi \rightarrow 0 \quad ,$$

and then

$$\Psi = \frac{c^2}{2} + \text{constant} \rightarrow 0 ,$$

or,

$$\Psi = \frac{g_{00}c^2}{2} - \frac{c^2}{2} \quad ,$$

and then

$$g_{00} = 1 + \frac{2\Psi}{c^2} . \qquad (3.6.4)$$

Therefore, g_{00} stands for a gravitational potential, and, consequently we justified in calling the metric tensor components as the "gravitational potentials". It is also clear from the intermediary step where we found,

$$\frac{d^2x^i}{ds^2} \approxeq -\Gamma^i_{00} \quad ,$$

that the Christoffel symbols represent accelerations, as we had advanced earlier.

2nd Condition: Obtainment of the value for Einstein's gravitational constant.

In the same conditions as above we now undertake the calculation of curvature tensor components R^i_{jkl}. As the Christoffel symbols are very small quantities, their products are negligible. So we can write:

$$R^i_{jkl} \approxeq \frac{\partial \Gamma^i_{jl}}{\partial x^k} - \frac{\partial \Gamma^i_{jk}}{\partial x^l} .$$

Therefore, we obtain Ricci tensor by:

$$R_{jk} \approxeq \frac{1}{2}\frac{\partial}{\partial x^k}\left[\frac{\partial \gamma_{ji}}{\partial x^i} + \frac{\partial \gamma_{ii}}{\partial x^j} - \frac{\partial \gamma_{ij}}{\partial x^i}\right] - \frac{1}{2}\frac{\partial}{\partial x^i}\left[\frac{\partial \gamma_{ij}}{\partial x^k} + \frac{\partial \gamma_{ki}}{\partial x^j} - \frac{\partial \gamma_{jk}}{\partial x^i}\right] =$$

$$\frac{1}{2}\left[\frac{\partial^2 \gamma_{ii}}{\partial x^j \partial x^k} + \frac{\partial^2 \gamma_{jk}}{\partial x^i \partial x^i} - \frac{\partial^2 \gamma_{ij}}{\partial x^i \partial x^k} - \frac{\partial^2 \gamma_{ki}}{\partial x^i \partial x^j}\right] .$$

In the particular case $j = k = 0$, we obtain:

$$R_{00} \approxeq \frac{1}{2}\left[\frac{\partial^2 \gamma_{ii}}{\partial x^0 \partial x^0} + \frac{\partial^2 \gamma_{00}}{\partial x^i \partial x^i} - 2\frac{\partial^2 \gamma_{10}}{\partial x^i \partial x^0}\right] .$$

In the static case,

$$R_{00} \approxeq \frac{1}{2}\frac{\partial^2 \gamma_{00}}{\partial x^i \partial x^i} \approxeq \frac{1}{2}\nabla^2 \gamma_{00} \approxeq \frac{1}{2}\nabla^2 g_{00} \quad .$$

In accordance with (3.6.4) the last expression reduces to:

$$R_{00} \approxeq \frac{1}{c^2}\nabla^2 \Psi \quad . \qquad (3.6.5)$$

We now are going to make a detour, and show that Einstein's field equations may be expressed, alternatively as

$$R_{ij} = \kappa \left[\frac{1}{2} g_{ij} T - T_{ij}\right] \quad , \qquad (3.6.6)$$

where

$$T \equiv T_i^i \quad .$$

By contracting the (3.4.3) field equation, we obtain:

$$G_i^i = \kappa T_i^i \quad \text{, or}$$

$$R = \kappa T \; . \qquad (3.6.7)$$

If we now plug (3.6.7) in equation (3.4.3), we obtain the desired relation (3.6.6).

After this detour, we return to (3.6.5) and we apply (3.6.6):

$$\frac{1}{c^2}\nabla^2 \Psi = \kappa \left(\frac{1}{2} g_{00} T - T_{00}\right) \quad . \qquad (3.6.8)$$

We now suppose the absence of other fields like for instance the electromagnetic; we suppose altogether that the matter distribution is that of "dust", i.e., $p \cong 0$, or,

$$T^{ij} = \rho u^i u^j \quad .$$

In the static case we shall find:

$$T_{00} \approxeq \rho \left(g_{00}\right)^2 \approxeq \rho \quad ,$$

and

$$T \approxeq \rho \left(g_{00}\right)^2 \approxeq \rho \quad .$$

On taking this values into (3.6.8), we obtain,

$$\frac{1}{c^2}\nabla^2 \Psi \approxeq -\frac{\kappa \rho}{2} \quad ,$$

or

$$\nabla^2 \Psi \approxeq -\frac{\kappa c^2}{2} \rho \; .$$

If we compare the above result with Poisson's equation (3.4.4):

$$\nabla^2 \Psi = -4\pi G \rho \ , \quad (3.4.4)$$

we obtain:

$$\boxed{\kappa = \frac{8\pi G}{c^2}} \quad (3.6.1)$$

3.7. The Geodesics Equation - a Derivation

We now derive, for the case of "dust" matter, immersed in a gravitational field, the equation of motion for each particle (each particle will follow a timelike geodesic).

For this case we found above that:

$$T^{ij} = \rho u^i u^j \ .$$

From the conservation of the covariant divergence of the energy-momentum tensor we have:

$$T^{ij}_{;j} = 0 \ .$$

It follows that:

$$\left(\rho u^j\right)_{;j} u^i + \rho u^j u^i_{;j} = 0 \ . \quad (3.7.1)$$

On multiplying the last equation per u^i , it follows that:

$$\left(\rho u^j\right)_{;j} u^i u_i + \rho u^i_{;j} u_i u^j = 0 \ . \quad (3.7.2)$$

From the definition of quadrivelocities we find:

$$u^2 = g_{ij} u^i u^j = g_{ij} \frac{dx^i}{ds} \frac{dx^j}{ds} = \frac{ds^2}{dsds} = 1 \ .$$

In other words,

$$u_i u^i = 1 \ . \quad (3.7.3)$$

On taking derivatives of both sides with respect to x^j, we find:

$$u_{i;j} u^i + u_i u^i_{;j} = 0 \ ,$$

or else,

$$u^i_{;j} u_i = 0 \ .$$

Taking this result to (3.7.2), we find:

$$\left(\rho u^j\right)_{;j} = 0 \ . \tag{3.7.4}$$

Taking this result to (3.7.1), we obtain:

$$u^j u^i_{;j} = 0 \ .$$

From the definition of covariant derivatives we have, for a contravariant vector,

$$u^i_{;j} \equiv \left[\frac{\partial u^i}{\partial x^j} + \Gamma^i_{kj} u^k\right] \ .$$

Therefore,

$$u^i_{;j} u^j \equiv \left[\frac{\partial u^i}{\partial x^j} + \Gamma^i_{kj} u^k\right] u^j = 0 \ .$$

Or,

$$\frac{d^2 x^i}{ds^2} + \Gamma^i_{kj} \frac{dx^j}{ds} \frac{dx^k}{ds} = 0 \ . \tag{3.7.5}$$

This is the famous geodesics equation, derived from Einstein's field equations, so that it has not to be adopted by means of an additional postulate within GRT. In the case of a particle not subject to the gravitational field, a straight line is our geodesic in Minkowski's space (Special Relativity) or in the three-dimensional Euclidean space (Newtonian Mechanics).

By employing variational calculus techniques, it is possible to demonstrate effectively that a geodesic equation in Riemannian geometry represents the curve, between two points of longest or shortest path. This is an additional evidence that the GRT reduces to Newtonian Mechanics where the counterpart of a geodesic is the straight line, known to be the shortest distance between two points in Euclidean space.

3.8. Gravitational Radiation

[We follow closely, here, the notation used by Kenyon (1990)].

We shall show that, in empty space, the linearized field equations, yield the wave equation. In empty space Einstein's equations become:

$$G_{\mu\nu} \equiv R_{\mu\nu} = 0 \tag{3.8.1}$$

Let us suppose that there is a small perturbation in the otherwise Minkowski's metric background:

$$g_{\mu\nu} = \eta_{\mu\nu} + h_{\mu\nu} \ , \tag{3.8.2}$$

where $\eta_{\mu\nu}$ stands for the Minkowski's metric coefficients, and the $h_{\mu\nu}$ are small perturbations. Notice that, the present linearized theory, differs from the previous one in Section 3.6 , which is more restrictive.

From the definition of the Christoffel symbols, we obtain:

$$\Gamma^{\alpha}_{\beta\delta} = \frac{1}{2} g^{\alpha\nu} [h_{\beta\nu,\delta} - h_{\delta\beta,\nu} + h_{\nu\delta,\beta}] \ . \qquad (3.8.3)$$

On the other hand, to first order in the perturbation, products of the $h_{\mu\nu}$'s are neglected and we find:

$$R_{\beta\delta} \cong \Gamma^{\alpha}_{\beta\delta,\alpha} - \Gamma^{\alpha}_{\beta\alpha,\delta} \cong \frac{1}{2} g^{\alpha\nu} [h_{\nu\delta,\beta\alpha} - h_{\delta\beta,\nu\alpha} + h_{\alpha\beta,\nu\delta} - h_{\alpha\nu,\beta\delta}] \cong 0 \ . \qquad (3.8.4)$$

(the zero result is due to the field equations).

As the $g^{\mu\nu}$'s also raise the subscripts of the $h_{\mu\nu}$'s, we find:

$$h^{\alpha}_{\delta,\beta\alpha} - h_{\delta\beta,\varepsilon\alpha} \eta^{\varepsilon\alpha} - h^{\alpha}_{\alpha,\beta\delta} + h_{\alpha\beta,\beta\delta} \eta^{\beta\alpha} \cong 0 \ . \qquad (3.8.5)$$

We may try to find a particular solution for the above equation. Suppose that:

$$h^{\alpha}_{\alpha} = 0 \ , \qquad (3.8.6)$$

and,

$$h^{\alpha}_{\delta,\alpha} = h_{\alpha\beta,\gamma} \eta^{\gamma\alpha} = 0 \ . \qquad (3.8.7)$$

Then we are left with:

$$h_{\delta\beta,\varepsilon\delta} \eta^{\varepsilon\alpha} \cong 0 \ . \qquad (3.8.8)$$

Equation (3.8.6) means that the perturbation is traceless; (3.8.7) means that in our particular solution we shall have zero divergence. Relation (3.8.8) is then:

$$\Box^2 h^{\mu\nu} = 0 \ , \qquad (3.8.9)$$

where $\Box^2$ stands for the d'Alembertian, i.e.:

$$\Box^2 \equiv -\eta^{\alpha\beta} \frac{\partial}{\partial x^{\alpha}} \frac{\partial}{\partial x^{\beta}} \equiv \frac{\partial^2}{\partial x^2} + \frac{\partial^2}{\partial y^2} + \frac{\partial^2}{\partial z^2} - \frac{1}{c^2} \frac{\partial^2}{\partial t^2} \equiv \nabla^2 - c^{-2} \frac{\partial^2}{\partial t^2} \ .$$

Equation (3.8.9) is the wave equation solution we were expecting. We make also the additional condition,

$$h_{\alpha 0} = 0 \ , \qquad (3.8.10)$$

and then, we can check that there is a freedom for choosing arbitrary only two independent components of $h_{\alpha\beta}$, which will be the two polarization states of the wave. The d'Alembertian equation means that the signal runs with the speed of light. In the Quantum outlook, the waves are carried by "gravitons", which are the equivalent of the photons in the electromagnetic case.

Gravitational waves were never been detected directly; nevertheless, there is indirect evidence of their existence, obtained by Hulse and Taylor who were awarded the Nobel Prize for their discovery (see Taylor and Weisberg, 1982; 1989). See Section 4.5 for a possible explanation of the negative results, so far, in attempted detections.

Chapter 4

Schwarzschild's Metric and Classical Experimental Tests

4.1. Spherically Symmetric Metrics

We define a metric with spherical symmetry around the origin, as a metric which is form-invariant according to the group of orthogonal transformations, of the type:

$$\widetilde{X} = AX \quad ,$$

where,

$$X \equiv (x, y, z) \quad ,$$

$$\widetilde{X} \equiv (\widetilde{x}, \widetilde{y}, \widetilde{z}) \quad ,$$

and

$$AA^T = I \quad ,$$

where I represents the identity matrix. (For the reader who needs to know, and does not, what are orthogonal transformations, next chapter has a review on it).

A metric is form-invariant if, in a transformation of coordinates $X \to \widetilde{X}$, we have,

$$g_{ij}(x) = \widetilde{g}_{ij}(\widetilde{x}) \quad .$$

Notice that the temporal coordinate is excluded from this spherical symmetry definition.

Form-invariants, for this group of transformations, are the following, up to second order in the coordinate differentials:

$$+x^2 + y^2 + z^2 \ ,$$

$$+xdx + ydy + zdz \ ,$$

$$+dx^2 + dy^2 + dz^2 \ .$$

In spherical coordinates (r, θ, ϕ), the above invariants become:

$$+r^2 ,$$

$$+rdr ,$$

$$+d\theta^2 + \sin^2\theta d\phi^2 .$$

We may also conclude, that the following are invariants: ($r, dr, d\theta^2, +\sin^2\theta d\phi^2$). In this way, the most general spherically symmetric metric in Riemann four-dimensional space (r, θ, ϕ, t) is:

$$ds^2 = A(r,t)dr^2 + B(r,t)(d\theta^2 + \sin^2\theta d\phi^2) + C(r,t)drdt + D(r,t)dt^2 , \qquad (4.1.1)$$

where the four coefficients A, B, C, D are up to now completely arbitrary.

Now let us remember Minkowski's metric:

$$ds^2 = -dx^\nu dx^\nu + c^2 dt^2. \qquad (\nu = 1, 2, 3)$$

In spherical coordinates, Minkowski's metric may therefore be expressed by:

$$ds^2 = -dr^2 - r^2(d\theta^2 + \sin^2\theta d\phi^2) + c^2 dt^2 . \qquad (4.1.2)$$

The metric which represents a mass distribution in the origin, should reduce, when the radial coordinate becomes very large ($r \to \infty$), to the Minkowski's metric; (in Cosmology, this may not be alike), so we take:

$$B(r,t) = r^2 .$$

In the case of a static gravitational field, the functions $A(r,t)$, $C(r,t)$, and $D(r,t)$ should not depend on the time coordinate t. On the other hand, the time dependence of the metric should be symmetric; i.e., the form of the metric can not be altered by a transformation $t \to -t$: we conclude that $C = 0$. We are left with the following metric:

$$ds^2 = -\left[e^{\alpha(r)}dr^2 + r^2(d\theta^2 + \sin^2\theta d\phi^2)\right] + e^{\beta(r)}c^2 dt^2 .$$

We choose in the above expression for the metric the exponentials $e^{\alpha(r)}$ and $e^{\beta(r)}$ so that they remain essentially positive and the signature of the metric should not change, at least for the time being.

We now begin the calculation of the metric tensor, the Christoffel symbols and the curvature and Ricci tensors, while attaching the following correspondence:

$$(x^0, x^1, x^2, x^3) \equiv (t, r, \theta, \phi) .$$

1st) Non-null covariant metric tensor components:

$$g_{11} = -e^{\alpha} \quad ,$$

$$g_{22} = -r^2 \quad ,$$

$$g_{33} = -r^2 \sin^2\theta \, ,$$

$$g_{00} = e^{\beta} c^2 \quad ,$$

$$g = det(g_{ij}) = -e^{(\alpha+\beta)} c^2 r^4 \sin^2\theta \, .$$

2nd) Non-null contravariant components of the metric tensor:

$$g^{11} = -\tfrac{1}{e^{\alpha}} = -e^{-\alpha} \, ,$$

$$g^{22} = -\tfrac{1}{r^2} \, ,$$

$$g^{33} = -\frac{1}{r^2 \sin^2\theta} \, ,$$

$$g^{00} = -\tfrac{1}{e^{\beta} c^2} .$$

3rd) Non-null Christoffel symbols of the second kind:

$$\Gamma^1_{11} = \tfrac{\alpha'}{2} \, ,$$

$$\Gamma^2_{12} = \Gamma^2_{21} = \tfrac{1}{r} \, ,$$

$$\Gamma^3_{13} = \Gamma^3_{31} = \tfrac{1}{r} \, ,$$

$$\Gamma^0_{10} = \Gamma^0_{01} = \tfrac{\beta'}{2} \, ,$$

$$\Gamma^1_{22} = -re^{-\alpha} \, ,$$

$$\Gamma^3_{23} = \Gamma^3_{32} = \cot g \; \theta \, ,$$

$$\Gamma^1_{33} = -re^{-\alpha} \sin^2\theta \, ,$$

$$\Gamma^2_{33} = -\sin\theta\cos\theta ,$$

$$\Gamma^1_{00} = -\tfrac{1}{2} c^2 \beta' e^{(\beta-\alpha)} \, .$$

(the primes indicate derivatives in relation to the radial coordinate).

4th) Ricci tensor non-null components:

$$R_{11} = \frac{\beta''}{2} + \frac{\beta'^2}{4} - \frac{\alpha'\beta'}{4} - \frac{\alpha'}{r} \ ,$$
$$R_{22} = e^{-\alpha}\left[r\frac{\beta'}{2} - \frac{r\alpha'}{2} + 1\right] - 1.$$

$$R_{33} = R_{22}\sin^2\theta \ .$$

$$R_{00} = c^2 e^{(\beta-\alpha)}\left[-\tfrac{1}{2}\beta'' - \tfrac{1}{4}\beta'^2 + \tfrac{1}{4}\alpha'\beta' - \tfrac{\beta'}{r}\right] .$$

Schwarzschild's metric exterior solution

The metric obtained by us, shall be certainly a solution for a static distribution, with spherical symmetry around the origin. For points that are exterior to the mass distribution in the origin, Einstein's field equations reduce to:

$$R_{ij} = 0 \ , \qquad (4.1.4)$$

because $T_{ij} = 0$.

According to the above calculation, we are left with the equations:

$$\beta'' + \tfrac{1}{2}\beta'^2 - \tfrac{1}{2}\alpha'\beta' - \tfrac{2}{r}\alpha' = 0 \ , \qquad (4.1.5)$$

$$\tfrac{1}{2}\beta' r - \tfrac{1}{2}\alpha' r + 1 = e^{\alpha} \ , \qquad (4.1.6)$$

$$\beta'' + \tfrac{1}{2}\beta'^2 - \tfrac{1}{2}\alpha'\beta' - \tfrac{2}{r}\beta' = 0 \ . \qquad (4.1.7)$$

Notice that we were left with three independent equations alone, because in our case,

$R_{22} = 0 \rightarrow R_{33} = 0$.

By comparing (4.1.5) and (4.1.7), we obtain:

$$\alpha + \beta = \text{constant.} \qquad (4.1.8)$$

The above equality should also be valid in spatial infinity where it should be identified by the values in the Minkowski's metric (4.1.2), and

$$e^{\alpha} = e^{\beta} = 1 \qquad \therefore \qquad \alpha \equiv \beta = 0 \ .$$

The constant in (4.1.8) is obviously zero-valued. And then, we find:

$$\alpha = -\beta \quad \text{at any place .} \qquad (4.1.9)$$

From (4.1.6) we now obtain

$$r\alpha' = 1 - e^{\alpha} \quad ,$$

or in other words,

$$\frac{dr}{r} = \frac{d\alpha}{1-e^{\alpha}} \quad .$$

When we integrate, the above equation, and we call "m" a constant of integration, we find:

$$e^{\alpha} = \frac{1}{1-\frac{2m}{r}} \quad . \qquad (4.1.10)$$

Then

$$e^{\beta} = 1 - \frac{2m}{r} \quad . \qquad (4.1.11)$$

We have thus obtained Schwarzschild's metric:

$$ds^2 = \frac{-dr^2}{1-\frac{2m}{r}} - r^2(d\theta^2 + \sin^2\theta d\phi^2) + c^2(1 - \frac{2m}{r})dt^2 \quad . \qquad (4.1.12)$$

The integration constant "m" can be given now a precise value: we remember that very far from the origin, where the mass M is located, the above metric can be identified with the metric of a weak field whose temporal metric coefficient is given by (3.5.4):

$$g_{00} = 1 + \frac{2\Psi}{c^2} \quad . \qquad (3.5.4)$$

We conjecture that the potential Ψ very far from the origin, is given by its Newtonian expression namely :

$$\Psi = -\frac{GM}{r} \quad . \qquad (4.1.13)$$

Obviously, we can identify the integration constant "m" with:

$$m = \frac{GM}{c^2} \quad . \qquad (4.1.14)$$

There is an apparent singularity of the Schwarzschild's metric for the point:

$$r = 2m = \frac{2GM}{c^2} \quad . \qquad (3.1.15)$$

This radial distance is called Schwarzschild's radius, and its value for the Earth, is about 9mm. If the Earth's mass would be entirely concentrated inside a sphere of radius smaller then 9mm, we would have a **black-hole**. By the so called process of gravitational collapse, its radius would become even more smaller than that, though I hope that it should not reduce to an infinitesimally small point of zero radius. Through the Schwarzschild's metric, a white-hole can also be envisaged with properties similar to a fountain of matter and photons surging from an apparent point-like source.

4.2. The Motion of Planets and Perihelion Precession

We shall now try to determine an equation for the orbital motion of a planet around the Sun. On considering valid the metric (4.1.12), we remember that each planet must obey approximately the path of a geodesic of the given space-time.

The equations of geodesics is given by (3.7.5), which we reproduce now:

$$\frac{d^2x^i}{ds^2} + \Gamma^i_{kj}\frac{dx^j}{ds}\frac{dx^k}{ds} = 0 \quad . \qquad (3.7.5)$$

With the values obtained for the Γ^i_{jk}'s in the prior section of this Chapter, we obtain the following equations:

$$\frac{d^2r}{ds^2} + \frac{\alpha'}{2}\left(\frac{dr}{ds}\right) - re^{-\alpha}\left(\frac{d\theta}{ds}\right)^2 - r\sin^2\theta e^{-\alpha}\left(\frac{d\phi}{ds}\right)^2 + \frac{e^{\beta-\alpha}}{2}\alpha'\left(\frac{dt}{ds}\right)^2 = 0 \quad , \qquad (4.2.1)$$

$$\frac{d^2\theta}{ds^2} + \frac{2}{r}\frac{dr}{ds}\frac{d\theta}{ds} - \sin\theta\cos\theta\left(\frac{d\phi}{ds}\right)^2 = 0 \quad , \qquad (4.2.2)$$

$$\frac{d^2\phi}{ds^2} + \frac{2}{r}\frac{dr}{ds}\frac{d\theta}{ds} - 2\cot\theta\frac{d\theta}{ds}\frac{d\phi}{ds} = 0 \, , \qquad (4.2.3)$$

$$\frac{d^2t}{ds^2} + \beta'\frac{dr}{ds}\frac{dt}{ds} = 0 \quad . \qquad (4.2.4)$$

Let us begin our study of the motion of the planet supposing that initially $\theta = \frac{\pi}{2}$ and $\frac{d\theta}{ds} = 0$. From (4.2.2) it follows that:

$$\frac{d^2\theta}{ds^2} = 0 \qquad \therefore \qquad \frac{d\theta}{ds} = \text{constant} = 0$$

(because zero was its initial value).

On the same token, $\theta = \frac{\pi}{2}$ = constant . (4.2.5)

The planet motion occurs always on the same plane, and is defined by equations (4.2.1), (4.2.3) and (4.2.4).

Now we turn to a solution of those equations. From the expression of the metric, we obtain an obvious solution:

$$1 + e^{\alpha}\left(\frac{dr}{ds}\right)^2 + r^2\left(\frac{d\phi}{ds}\right)^2 - e^{\beta}\left(\frac{dt}{ds}\right) = 0 \quad . \qquad (4.2.6)$$

Equation (4.2.3) becomes now:

$$\frac{du}{dr}\frac{dr}{ds} + \frac{2}{r}\frac{dr}{ds}u = 0 \quad ,$$

where we define $u \equiv \frac{d\phi}{ds}$.

A trivial solution would be r =constant. If we drop it, we are left with:

$$\frac{du}{u} = 2\frac{dr}{r} \ ,$$

If C is a constant of integration, we find from the above:

$$u = \frac{d\phi}{ds} = \frac{C}{r^2} \ . \qquad (4.2.7)$$

If we now define $n \equiv \frac{dt}{ds}$, equation (4.2.4) becomes:

$$\frac{dn}{d\beta} + n(\beta) = 0 \ ,$$

By integration, we find:

$$n = n(\beta) = \frac{dt}{ds} = Le^{-\beta} \ , \qquad (4.2.8)$$

where L is a constant of integration.

We now take (4.2.8) to (4.2.6), and on remembering the values of α and β which we obtained in the prior section, we find:

$$1 + \frac{m}{r} + \left(\frac{dr}{ds}\right)^2 + r^2\left(\frac{d\phi}{ds}\right)^2\left(1 + \frac{m}{r}\right) - L^2 = 0 \ . \qquad (4.2.9)$$

Let us define a new constant:

$$h = L^2 - 1 \ .$$

Equation (4.2.9) may now be written as:

$$\left(\frac{dr}{ds}\right)^2 + r^2\left(\frac{d\phi}{ds}\right)^2 + \frac{m}{r}\left[1 + r^2\left(\frac{d\phi}{ds}\right)^2\right] = h \ , \qquad (4.2.10)$$

while from (4.2.7) it follows that:

$$r^2\left(\frac{d\phi}{ds}\right)^2 = C \ . \qquad (4.2.11)$$

Equations (4.2.5), (4.2.10), and (4.2.11) rule the motion of a planet according to GRT. In order to interpret the above solution we turn our eyes to the corresponding solution in Newtonian Mechanics whose three laws are:

1st Law: Central motion is contained on a plane.

$\therefore \ \theta = \frac{\pi}{2}$ is a solution.

2nd Law: Angular momentum is conserved.

$$\therefore \ r^2\frac{d\phi}{dt} = C' = \text{constant.} \qquad (4.2.12)$$

3rd Law: The sum of Kinetic plus potential energy is conserved.

$$\left(\frac{dr}{dt}\right)^2 + r^2\left(\frac{d\phi}{dt}\right)^2 - \frac{2GM}{r} = h' = \text{constant.} \quad (4.2.13)$$

For the comparison between GRT and Newtonian Mechanics results to be effective, we remember that in Newtonian Mechanics the proper time confounds itself with time "t" so that for the Newtonian metric we can write:

$$ds = cdt \ . \quad (4.2.14)$$

On a comparison we identify:

$$m = \frac{GM}{c^2}.$$

This result confirms the derivation done in section 4.1 (Formula (4.1.14)).

In Newtonian Mechanics, the solution for (4.2.12) and (4.2.13) is straightforward:

$$\frac{d^2u}{d\phi^2} + u = \frac{GM}{C'^2}\ , \quad (4.2.15)$$

where

$$u = \frac{1}{r}.$$

The Newtonian solution is a conic, given by:

$$u_N = \frac{1}{r} = \frac{GM}{C'^2}\left[1 + e\cos\phi\right], \quad (4.2.16)$$

where e stands for the eccentricity of the conic.

We now return to the relativistic equations which for our purpose of comparison can be written as:

$$\frac{dr}{ds} = \frac{dr}{d\phi}\frac{d\phi}{ds} = \frac{dr}{d\phi}\frac{C}{r^2} = -C\frac{du}{d\phi}\ , \quad (4.2.17)$$

and,

$$\left(\frac{du}{d\phi}\right)^2 + 2u\frac{du}{d\phi} - \frac{2m}{C^2}\frac{du}{d\phi} - 6mu^2\frac{du}{d\phi} = 0\ . \quad (4.2.18)$$

Differentiating the last equation in relation to ϕ , and applying equation (4.2.17), we obtain:

$$\frac{d^2u}{d\phi^2} + u = \frac{m}{C^2} + 3mu^2 \quad . \quad (4.2.19)$$

On comparison with (4.2.14) we identify:

$$\frac{m}{C^2} \equiv \frac{GM}{C'^2}\ .$$

The additional relativistic term,

$$3mu^2 = \frac{3GM}{c^2}\frac{1}{r^2} \quad ,$$

turns out to be much smaller than the other terms, so it is a "perturbative" term.

We know then, that the relativistic orbit of the planet can only have a small deviation from an ellipse which is the Newtonian basic solution. In the perturbative term, we plug tentatively the Newtonian solution for the ellipse, u_N , so we write:

$$\frac{d^2u}{d\phi^2} + u \cong \frac{m}{C^2} + 3m(u_N)^2 \quad . \qquad (4.2.20)$$

It is evident that the solution of the above equation must render a better approximation than the solution:

$$u \cong u_N \ ,$$

for the equation (4.2.20).

We now substitute u_N by the value given in (4.2.16) and plug into (4.2.20); we make the approximation $e^2 << 1$, and find :

$$\frac{d^2u}{d\phi^2} + u = \frac{m}{C^2} + \frac{6m^3}{C^4} e \cos\phi, \qquad (4.2.21)$$

where we also considered:

$$\frac{3m^2}{C^2} << 1 \ .$$

The solution to the above equation is:

$$u = \frac{1}{r} = \frac{m}{C^2}\left[1 + e\cos\left(\phi - \frac{3m^2}{C^2}\phi\right)\right] \quad . \qquad (4.2.22)$$

On comparison between (4.2.22) and (4.2.16), we verify that there is an advance in the principal axis of the ellipse, which for a round turn ($\Delta\phi = 2\pi$) is given by:

$$\Delta\omega = \frac{6\pi m^2}{C^2} \quad .$$

On the other hand, if we call "a" the semi major axis of the ellipse,

$$\frac{C^2}{m} = a(1-e^2) \quad .$$

This is tantamount to:

$$\therefore \quad \boxed{\Delta\omega = \frac{6\pi GM}{c^2 a(1-e^2)}} \quad . \qquad (4.2.23)$$

For the planet Mercury,

$$\Delta w \simeq 0,104''.$$

In one century the calculated advance would be 43,0" as compared to the experimental value, from astronomical observations, 43,1".

We can not forget that the total advance in the perihelion of the planet Mercury is 5,600". The non-relativistic perturbations accounted by the interference of other planets' orbits accounts for 5,557" and only 43,41" are explained by GRT corrections, not justified in Newtonian Mechanics. This match was perhaps the most astonishing result of Albert Einstein's theory, and testifies in favor of GRT.

4.3. Propagation of Light near Gravitational Fields

In Special Relativity Theory the light trajectory is given by:

$$ds = 0 \quad .$$

Though this result is a bit obvious we shall demonstrate it for the sake of completeness, and because many beginning students do not find it explicit from some textbooks.

We recall Minkowski's metric:

$$ds^2 = -\left[dx^2 + dy^2 + dz^2\right] + c^2 dt^2 \quad . \qquad (2.1.4)$$

If the speed of a particle is defined kinematically as:

$$V = \sqrt{dx^2 + dy^2 + dz^2}/dt \ ,$$

and we apply this formula for a photon, we know that the result is $V = c$. Plugging back into (2.1.4) we find $ds = 0$. A result valid for photons and thus for light rays. When we go to GRT we are led to keep the equation for light rays as:

$$ds = 0 \ . \qquad (4.3.1)$$

For a photon in the neighborhood of a central mass M located at the origin of the co-ordinate system with a spherically symmetric distribution, we take Schwarzschild's metric, and the equations (4.2.19) and (4.2.11) can be kept along with (4.3.1). In our case the value of the constant C which appears in (4.2.11) is to be taken, obviously, as infinite:

$$C = \infty. \qquad (4.3.2)$$

This being the case we are left from (4.2.19) with:

$$\frac{d^2u}{d\phi^2} + u = 3mu^2 \cdot \qquad (4.3.3)$$

We take into account that the r.h.s. of (4.3.3) represents a small corrective or perturbative term, in the case of the Sun for example. In that case, $u = \frac{1}{r}$ would be calculated with r equal to the Sun's radius, and it is clear that the approximate solution of this equation would be $u \cong u_1$, where u_1 stands in the equation:

$$\frac{d^2u_1}{d\phi^2} + u_1 \cong 0 \ .$$

Its solution is:

$$u_1 = \frac{\cos\phi}{R},$$

where R stands for a constant.

A better approximation may now be obtainable by solving the following equation:

$$\frac{d^2u}{d\phi^2} + u = 3m(u_1)^2 = \frac{3m\cos^2\phi}{R^2} \qquad (4.3.4)$$

The reader should pay attention and focus himself that the above equation (4.3.4) comes in fact from (4.3.3) where in r.h.s. we have substituted u by its Newtonian solution.

The solution of (4.3.4) is

$$u = \frac{1}{R}\cos\phi + \frac{m}{R^2}\left(2 - \cos^2\phi\right).$$

The equation which one finds for the deviation of light rays in the neighborhood of a central gravitational mass like the Sun, is obtainable if we compare the two farthest rays for which:

$$u = \frac{1}{r} = 0\,.$$

We have an equation of the second degree in $\cos\phi$:

$$(m/R)\cos^2\phi - \cos\phi - (2m/R) = 0.$$

Its roots are:

$$\cos\phi = \left[1 \pm \sqrt{1 + \frac{8m^2}{R^2}}\right](2m/R)^{-1}\,.$$

In our approximation:

$$\frac{m^2}{R^2} << 1\,.$$

We now find:

$$\cos\phi \cong \left[1 \pm \left(1 + \frac{4m^2}{R^2}\right)\right](2m/R)^{-1}\,.$$

The first solution is:

$$\cos\phi \cong (2m/R)\,.$$

The second solution is:

$$\cos\phi' \cong \left(2 + \frac{4m^2}{R^2}\right)(2m/R)^{-1} = \frac{R}{m} + \frac{2m}{R}\,.$$

The physically acceptable solution is the first one, because in this solution the deflection of light will be small accordingly to what one expects when we employ GRT instead of Special Relativity. This confirms that GRT offers small corrections to both S.R. and Newtonian Mechanics when we are not facing extremely powerful gravitational fields like for instance it would happen in the case of the gravitational collapse of an astrophysical object into a black-hole.

As in S.R. we would find $\cos\phi = 0$, we take now the solution:

$$\cos\phi << 1 \ .$$

The final result is:

$$\phi \cong \pm\left(\frac{\pi}{2} + \frac{2m}{R}\right) .$$

This angle is valid for the two farthest extremes of the light path, bringing a total deviation of:

$$\Delta\phi = \frac{4m}{R} . \qquad (4.3.5)$$

The predicted theoretical deviation for the Sun, with a light ray passing close to its exterior, is thus:

$$\Delta\phi = 1.75'' .$$

We may compare the above theoretical value with the experimental measurement for the Quasar 3C279 which is blackened by the Sun once in a year: the radio waves emitted by this object showed a deviation:

$$\Delta\phi = 1.73'' \pm 0.05'' .$$

The predicted value is 97% accurate !!!

4.4. Spectral Displacement due to Gravitational Cause

From a star of gravitational mass M and radius R_1 a certain electromagnetic radiation is emitted. Consider an observer located at a distance R_2 from the origin of the coordinate system which coincides with the center of the star, which we suppose homogeneous and spherical. We shall suppose that $R_2 >> R_1$. The Schwarzschild's metric is applicable:

$$ds^2 = \frac{-dr^2}{1-\frac{2m}{r}} - r^2 d\theta^2 - r^2 \sin^2\theta d\phi^2 + (1 - \frac{2m}{r})dt^2 . \qquad (4.1.12)$$

Throughout this book we have not dealt with proper time τ up to now. It should not be identified with the variable "t" which stands for coordinate time. We define the former by:

$$d\tau = \frac{ds}{c},$$

with

$$dr = d\theta = d\phi = 0 .$$

Accordingly we shall have:

$$d\tau = \frac{ds}{c} = \sqrt{1 - \frac{2m}{R}} dt,$$

(with $dr = d\theta = d\phi = 0$).

Let us suppose that at coordinate times t_1 and $t_1 + dt_1$, two successive wave crests are emitted from the surface of the star. The proper period of emission $d\tau_1$ is given by:

$$d\tau_1 = \sqrt{1 - \frac{2m}{R_1}} dt_1 . \qquad (4.4.1)$$

If the observer, say, on the Earth's surface receives the impulses with a time difference dt_2 , the proper period of reception $d\tau_2$ is given by:

$$d\tau_2 = \sqrt{1 - \tfrac{2m}{R_2}} dt_2 \ . \qquad (4.4.2)$$

We are in face of a certain sort of "comoving" observers, which for the time being we define as observers static in relation to the source. We further consider that the period dt_2 $<< 1$, and so, we commit a forgivable mistake if we suppose that:

$$dt_1 \cong dt_2 \ . \qquad (4.4.3)$$

If λ stands for the emitted wavelength, and $(\lambda + \Delta\lambda)$ is the received corresponding value,

we shall have:

$$\tfrac{(\lambda+\Delta\lambda)}{\lambda} = 1 + z = (1 - \tfrac{2m}{R_2})^{1/2}(1 - \tfrac{2m}{R_1})^{-1/2} \cong (1 - \tfrac{2m}{R_1})^{-1/2} \cong 1 + \tfrac{m}{R_1} \ . \qquad (4.4.4)$$

In the year 1965, Pound and Snyder made a precise measurement of a gravitational redshift in the Earth's gravitational field, by means of the Mössbauer effect. This experiment has been repeated with verification of formula (4.4.4) with 99% of confidence (see Pound and Rebka (1960)).

4.5. Other Verifications of the GRT

1st) Gravitomagnetism and galactic rotations

GRT has been verified exhaustively for about 100 years and no discrepancies have been found up to now. Gravitomagnetic phenomena were predicted for a long time until, Ignazio Ciufolini and others (Berman,2007), reported recently that satellites carrying gyroscopes had verified closely the approximate solution (Lense-Thirring metric) given by the Austrian scientists Lense and Thirring at the beginning of 20th century. It turns out that in the same way as charged moving particles create a magnetic field, rotating masses create an additional "gravitomagnetic effect" . Likewise, Cooperstock and Tieu (2005) solved the galactic rotational problem, by means of GRT, finding that there is no use in imposing the existence of exotic "dark" matter, in order to explain the high tangential speeds of orbiting stars around, and far from, the center of the galaxy (Berman,2007).

2nd) Hafele-Keating experiment:

In section 4.4 we mentioned the relation between proper time $d\tau_1$ and coordinate time dt for Schwarzschild's metric:

$$d\tau_1 = \sqrt{1 - \tfrac{2m}{R}} dt \ , \qquad (4.5.1)$$

in the presence of a gravitational field. On the other hand, in Special Relativity there is a second variation between proper time $d\tau_2$ and the coordinate time dt for a moving object with speed v :

$$d\tau_2 = \sqrt{1 - v^2/c^2}dt \, . \tag{4.5.2}$$

Equation (4.5.1) tells us that clocks go slower in the presence of strong gravitational fields. Relation (4.5.2) tells us that clocks go faster when they move, relative to static clocks. By means of Hafele-Keating experiment it was possible to compare clocks readings inside airplanes with similar clocks on the Earth's surface. A team of the University of Maryland in the United States, made measurements between the years 1975 and 1976. Due to differences in the intensity of the gravitational fields on the Earth's surface, and the flying airplanes it was found an advancement in time measured by the clocks inside airplanes according to (4.5.2). However, this effect was partially camouflaged by the opposite effect due to the speed of the airplanes relative to static clocks on the Earth's surface. By employing slow motion airplanes, this opposite effect was reduced to less than 10% of the total retardation; thus formula (4.5.1) could be verified experimentally with 98% of precision relative to the theoretical value.

3rd) Shapiro experiment:

In the year 1965, the American scientist I. I. Shapiro proposed the following experiment: a radar wave emission should be directed to another planet so that the light ray would pass near the Sun. The planet would reflect the radar emission, the time lag would be measured and obviously the time taken would be larger than expected due to the passage near the Sun. In the year 1970 Shapiro made the measurement confirming Einstein's theory with 97% of accuracy (Adler et al., 1975).

4th) Hulse-Taylor observations:

As we mentioned in last Chapter, GRT's predictions of the existence of gravitational waves, got indirect confirmation by means of the observation of a Pulsar, with a possible companion; they would loose energy due to gravitational radiation according to the general relativistic calculation (Einstein's formula), not derived or shown in this book.

Note on gravitational radiation: it is my opinion that gravitational waves' direct detection, may be hindered by a misleading interpretation of the tensors formulae; there is a great difference between "tensor components", and "physical tensor components". The latter, are the ones that are experimentally measured, and they may be affecting the order of magnitude of the sought slight relative displacement on the detectors. (Synge and Schild, 1969).

5th) Global Positioning System (GPS) — NAVSTAR :

By continuously emitting radio waves, at fixed frequencies, several orbiting satellites can activate a GPS detector, (for instance, inside an airplane), which arrives with a Doppler shift, dependent on the relative speed of the source and the receiver; it also depends on the gravitational potential difference between the locations of the source and receiver. Both Special and General Relativity are involved; speeds and positions of the airplanes are then measured with extreme accuracy ($\pm$ 2 cm.s^{-1} ; 16 m) . This system of navigation is called GPS-NAVSTAR, and works well (Halliday et al.,2005).

Final Comments

Einstein has commented that it would suffice a single negative result breaking GRT's predictions, in order that this theory be discarded, despite all other confirmatory results.

Chapter 5

Complements of Tensor Calculus and General Relativity (Optional Study)

5.1. Orthogonal Transformations and Cartesian Tensors

Two coordinate systems S and S' relate themselves by means of an orthogonal transformation, if it can be expressed like:

$$x'^i = a_{ij}x^j + b^i \ , \qquad (5.1.1)$$

where a_{ij} and b^i are constants.

In a orthogonal transformation, the distance between two points, x_0^i and x^i is defined by its square:

$$s'^2 = (x'^i - x_0'^i)(x'^i - x_0'^i),$$

which is an invariant, i.e.:

$$s'^2 = s^2 \ . \qquad (5.1.2)$$

We define:

$$\Delta x^i = x^i - x_0^i \ ,$$

and we verify that (5.1.1) can be written like:

$$\Delta x'^i = a_{ij}\Delta x^j \ , \qquad (5.1.3)$$

or in matrix form:

$$\Delta x' = A\Delta x \ . \qquad (5.1.4)$$

If we call the transpose of $\Delta x'$, as $(\Delta x')^T$, we may write:

$$(\Delta x')^T(\Delta x') = (\Delta x)^T(\Delta x) = s^2 \ . \qquad (5.1.5)$$

On taking the transpose of both sides of (5.1.4) we find:

$$(\Delta x')^T = (\Delta x)^T A^T \quad . \tag{5.1.6}$$

We now combine relations (5.1.4), (5.1.5) and (5.1.6), finding:

$$(\Delta x)^T A^T A(\Delta x) = (\Delta x)^T (\Delta x). \tag{5.1.7}$$

From the above equation, we find:

$$A^T A = I \quad , \tag{5.1.8}$$

where the symbol I stands for the identity matrix.

By taking determinants of both sides of the last equation, we find:

$$(\det A)^2 = 1 \quad ,$$

because,

$$\det A^T = \det A \; .$$

We conclude that :

$$\det A \;=\; \pm 1 \; . \tag{5.1.9}$$

From (5.1.8) it results also that:

$$A^T = A^{-1} , \tag{5.1.10}$$

where A^{-1} stands for the inverse of the matrix A.

The coefficients a_{ij} of the matrix A, in consequence, obey to relation:

$$a_{ji} a_{jk} = \delta_{ik} \; , \tag{5.1. 11}$$

due to relation (5.1.8). They also obey to the relation:

$$a_{ji} a_{ki} = \delta_{jk} \; , \tag{5.1. 12}$$

due to obvious relation:

$$AA^T = I \; ,$$

which is obtained from (5.1.10) by multiplying both sides of the equation with the matrix A:

$$AA^T = AA^{-1} = I.$$

Relation (5.1.11) and (5.1.12) are the necessary conditions for the transformation to be orthogonal, and each one of them is a sufficient condition for this transformation.

Cartesian Tensors

The definition of tensors given on Chapter 1, can be simplified for orthogonal transformation by means of the relation:

$$\frac{\partial x'^i}{\partial x^j} = a_{ij} \ . \qquad (5.1.13)$$

The above was obtained from (5.1.1). Notice that there is no distinction among covariant, mixed or contravariant components, because:

$$\frac{\partial x^i}{\partial x'^j} = \frac{\partial x'^i}{\partial x^j} = a_{ij} \ . \qquad (5.1.14)$$

5.2. Parallel Transport in Riemann Space

We have been defining tensors and operations with tensors over all space. Nevertheless, in some occasions, it is convenient to deal with a vector $A^i(s)$ defined only on a curve $x^k = x^k(s)$, where "s" is a curve parameter, that we shall preferably associate with the arc length. Obviously there is no sense in speaking on its covariant derivative, $\frac{DA^i}{Dx^k}$ over all space, relative to a variable x^k . Nevertheless, we can talk of the covariant, or "absolute", derivative of the vector along the curve of parameter "s" which we shall denote by $\frac{DA^i}{Ds}$.

By definition,

$$A'^i(s) = \frac{\partial x'^i}{\partial x^j} A^j(s) \ . \qquad (5.2.1)$$

By the rules of differentiation, we have:

$$\frac{dA'^i}{ds} = \frac{\partial x'^i}{\partial x^j}\frac{dA^j(s)}{ds} + \frac{\partial^2 x'^i}{\partial x^k \partial x^j} A^j(s) \frac{dx^k}{ds} \ . \qquad (5.2.2)$$

By analogy with the study already done, for the definition of covariant derivative, in Riemann space, it is clear that we may define along a given curve, the absolute derivative as follows:

$$\frac{DA^i}{Ds} \equiv \frac{dA^i}{ds} + \Gamma^i_{jk}\frac{dx^k}{ds} A^j \ . \qquad (5.2.3)$$

Analogously, we could define covariant derivatives or absolute derivatives along a given curve, for any kind of tensor. For any kind of tensor, say T^{ij}_k we may define, by formal manipulation a relation between the absolute derivative along a curve and its covariant derivative over all space, even if it does not exist in practice:

$$\frac{DT^{ij}_k}{Ds} \equiv \frac{DT^{ij}_k}{Dx^m} \cdot \frac{dx^m}{ds} \ .$$

It must be remembered, however, that the covariant derivative along a curve may be the only one that we may calculate.

We define now the **parallel transport** of a tensor along a given curve, as the law which makes that its covariant derivative along the curve be zero. For instance, consider a contravariant vector A^i . The law of its parallel transport is given by:

$$dA^i = -\Gamma^i_{jk} A^j dx^k \ . \qquad (5.2.4)$$

In order to understand the physical meaning of this law, consider parallel transport in a geodesic system of coordinates:

$$dA^i = 0 \ .$$

Obviously its covariant derivative is also null:

$$\frac{DA^i}{Ds} = 0 \ .$$

(in this coordinate system)

Thus, the vector does not change its size or direction. As the tensor properties are independent of the particular coordinate system to be employed, what we have said, is valid for the general case: the transported vector "does not change its size or direction". By means of parallel transport, we may compare two tensors located at different points of spacetime.

We may now interpret anew the equations of geodesics. As we saw in Chapter 3, it is given by:

$$\frac{d^2x^i}{ds^2} + \Gamma^i_{jk} \frac{dx^j}{ds} \frac{dx^k}{ds} = 0 \ .$$

We rewrite the above as:

$$\frac{D}{Ds}\left(\frac{dx^i}{ds}\right) = 0 \ . \qquad (5.2.5)$$

Or:

$$\frac{D}{Ds} t^i = 0 \ , \qquad (5.2.6)$$

where $t^i = \frac{dx^i}{ds}$ is the vector tangent to the curve $x^i = x^i(s)$ by definition.

We are now in condition to say that the geodesics equation warrants that the tangent vector to geodesic curve is parallel-to-itself transported along the curve.

5.3. Fermi-Walker's Transport

When the observer does not move along a geodesic, the tangent vector $t^i = \frac{dx^i}{ds}$ will not coincide at another point of the curve with its paralleled transported vector. There is another transport law, the Fermi-Walker's, that assures that the Fermi-derivative:

$$\frac{D_F A^i}{Ds} = \frac{DA^i}{Ds} - A_j \left[\frac{dx^i}{ds} \frac{D^2 x^j}{Ds^2} - \frac{dx^j}{ds} \frac{D^2 x^i}{Ds^2} \right], \tag{5.3.1}$$

is zero along the curve.

If the curve is a geodesic, $\frac{D_F}{Ds}$ coincides with $\frac{D}{Ds}$: Fermi-Walker's transport law for a geodesic coincides with parallel transport.

In order to show that the tangent vector to a curve transport itself along Fermi-Walker, the reader can check that, in (5.3.1) with:

$$A^i = \frac{dx^i}{ds},$$

we obtain:

$$\frac{D_F}{Ds}\left(\frac{dx^i}{ds}\right) = 0 .$$

On the other hand, write $\frac{D^2 x^n}{Ds^2} = 0$ in (5.3.1) in order to obtain:

$$\frac{D_F}{Ds}\left(\frac{dx^n}{ds}\right) = 0 .$$

In other words, if a particle moves along a geodesic, the laws of parallel transport and Fermi-Walker's transport, coincide. We interpret (5.3.1) as follows: Fermi-Walker's transport law allows one to measure the variation of a vector A^i along a curve, departing from an initial point. It can be shown that the scalar product of vectors given by the expression $g_{ij}A^iA^j = A_jA^j$ is not altered by Fermi-Walker's transport, so that, lengths and vectors are preserved (Stephani,1990).

5.4. Lie Derivative

We now define a new type of tensor derivative called the Lie derivative (Carmeli,1982). Consider an infinitesimal transformation of coordinates:

$$x'^i = x^i + \varepsilon a^i(x) , \tag{5.4.1}$$

where ε is an infinitesimal parameter and $a^i(x)$ is a contravariant vector field which may be defined by:

$$a^i(x) = \left(\frac{\partial x'^i}{\partial \varepsilon}\right)_{\varepsilon=0} \tag{5.4.2}$$

Let now $T(x)$ be any tensor of any order, defined in each point of spacetime. Denote by P and Q two neighboring points with coordinates (x^i) and $(x^i + \varepsilon a^i)$ respectively. In point Q , the tensor field T may be calculated by two distinct modes:

1st mode: $T = T(\, x'^i)$. In this mode we have a value given by the definition of the field in each point including point Q.

2nd mode: $T' = T'(\, x'^i)$. In this mode we obtain $T'(x'^i)$ by means of $T(x^i)$ by application of definition of a tensor transformation from system of coordinates $x \to x'$, i.e. , $P \to Q$.

We now define:

$$L_{a^i} T(x) = \lim_{\varepsilon \to 0} \left[\frac{T(x') - T'(x')}{\varepsilon} \right], \qquad (5.4.3)$$

and call this, the Lie derivative of tensor $T(x)$ in the direction of vector $a^i(x)$, at point Q.

Alternatively, because points P and Q coincide in the limit, we could have defined the same Lie derivative as above by means of:

$$L_{a^i} T(x) = \lim_{\varepsilon \to 0} \left[\frac{T(x) - T'(x)}{\varepsilon} \right] . \qquad (5.4.4)$$

The two definitions are equivalent; the Lie derivative measure the variation of a tensor when the observer moves from one point to a neighboring point in a given direction, retaining its coordinate system.

Example 1: Calculate $L_a \phi(x)$, where ϕ is an invariant,

Consider that:

$$\phi(x) = \phi'(x'). \qquad (5.4.5)$$

Let us apply the first definition: on point Q we have:

$$T(x') = \phi(x') = \phi(x + \varepsilon a) = \phi(x) + \varepsilon a^i \frac{\partial \phi}{\partial x^i} ,$$

and,

$$T'(x') = \phi'(x') = \phi(x) .$$

Then,

$$L_a \phi(x) = \lim_{\varepsilon \to 0} \left[\phi(x) + \varepsilon a^i \frac{\partial \phi}{\partial x^i} - \phi(x) \right] \varepsilon^{-1} = a^i \frac{\partial \phi}{\partial x^i} . \qquad (5.4.6)$$

Example 2: Calculate $L_a V^i(x)$, i.e., obtain the Lie derivative of a contravariant vector.

Let us apply the second definition:

$$V'^j(x') = V'^j(x'^i + \varepsilon a^i) = V'^j(x) + \varepsilon a^i \frac{\partial V'^j}{\partial x^i} \cong V'^j + \varepsilon a^i \frac{\partial V^j}{\partial x^i} \,, \qquad (5.4.7)$$

where in the last term, we neglected second order infinitesimals ($\varepsilon^2 \cong 0$). From the definition of a contravariant transformation,

$$V'^j(x') = V^k(x) \frac{\partial x'^j}{\partial x^k} \,. \qquad (5.4.8)$$

On the other hand,

$$\frac{\partial x'^j}{\partial x^k} = \frac{\partial}{\partial x^k}\left[x^j + \varepsilon a^j\right] = \delta^j_k + \varepsilon \frac{\partial a^j}{\partial x^k} \,. \qquad (5.4.9)$$

From the last two relations, we obtain:

$$V'^j(x') = V^k \delta^j_k + \varepsilon V^k \frac{\partial a^j}{\partial x^k} = V^j + \varepsilon V^k \frac{\partial a^j}{\partial x^k} \,. \qquad (5.4.10)$$

From (5.4.7) and (5.4.10), we find:

$$V'^j + \varepsilon a^i \frac{\partial V^j}{\partial x^i} = V^j + \varepsilon V^k \frac{\partial a^j}{\partial x^k} \,.$$

Then,

$$L_a V^i(x) = \lim_{\varepsilon \to 0}\left[\varepsilon\left(a^i \frac{\partial V^j}{\partial x^i} - V^k \frac{\partial a^j}{\partial x^k}\right)\right]\varepsilon^{-1} = a^i \frac{\partial V^j}{\partial x^i} - V^i \frac{\partial a^j}{\partial x^i} \,. \qquad (5.4.11)$$

Analogously, we could find the Lie derivatives of any other kind of tensor.

Important result: in a Riemann space, the reader might check that an alternative but equivalent expression for relation (5.4.11) could be obtained by substituting partial derivatives by covariant derivatives; this conclusion is general and allows us to infer that the Lie derivative of a tensor, in Riemann space, is also a tensor.

Further results.

1st) The reader may check that:

$$L_a V_n = V_{n;i} a^i + V_i a^i_{;n} \,.$$

$$L_a g_{mn} = a_{m;n} + a_{n;m} \,, \qquad (5.4.12)$$

where g_{mn} is the metric tensor component.

2nd) it obeys the usual rules for the derivative of sums and products.

3rd) it commutes with the contraction.

4th) it commutes with partial derivatives.

5.5. Isometries

The study of the symmetries in Riemann space, begins with the calculation of the variation of the metric tensor in given directions of space.

Definition: Isometric mapping is an infinitesimal coordinate transformation under which the Lie derivative of the metric tensor becomes null:

$$L_a g_{ij}(x) = 0 \,. \qquad (5.5.1)$$

On remembering relation (5.4.12), we find then:

$$a_{i;j} + a_{j;i} = 0 \,. \qquad (5.5.2)$$

This is the famous Killing equation, whose solutions, if they exist, allow the isometric mapping. The solutions $a_i(x)$ are the Killing vectors, and if they exist, we say that the space has intrinsic symmetries.

Example: Find the intrinsic symmetries of the Euclidean plane, i.e., solve Killing's equations for the Euclidean plane.

Solution:

In Cartesian coordinates,

$$ds^2 = dx^2 + dy^2$$

$$\therefore g_{AB} = \delta_{AB} \quad \text{where} \quad A, B = 1, 2 \,.$$

Killing equation reduces to:

$$\frac{\partial a^1}{\partial x} = \frac{\partial a^2}{\partial y} = 0 \,. \qquad (5.5.3)$$

and,

$$\frac{\partial a^1}{\partial y} + \frac{\partial a^2}{\partial x} = 0 \,. \qquad (5.5.4)$$

By integration of the first equation (5.5.3) we find:

$$a^1 = Y(y) \,.$$

and,

$$a^2 = X(x) \,.$$

On taking to the second equation (5.5.4) the above results become:

$$\frac{dY(y)}{dy} + \frac{dX(x)}{dx} = 0 \,.$$

This last equation can be solved by the method of separation of variables:

$$\frac{dY(y)}{dy} = -\phi \; = \text{constant.}$$
$$\frac{dX(x)}{dx} = +\phi \; = \text{constant.}$$

With the following solutions:

$$Y = -\phi y + x_0 \,. \qquad (5.5.5)$$
$$X = +\phi x + y_0 \,,$$

where (x_0, y_0) is a pair of integration constants.

Geometrical interpretation of the above result:

We have three degrees of freedom in the solution, which describe the infinitesimal group of motions of the bidimensional Euclidean space; these are the translations x_0 and y_0 along the axes OX and OY, and rotations around the origin $x = y = 0$ according to an angle arc sin ϕ.

5.6. Stationary and Static Fields

A gravitational field is said to be stationary if it admits a Killing temporal vector, i.e., $a_i a^i > 0$. From the Killing equation:

$$a_{i;j} + a_{j;i} = 0 \,,$$

where,

$$a_i = g_{ij} a^j \,.$$

We shall now investigate the above definition. We build a coordinate system such that only the coordinate x^0 varies along the trajectories of a^i keeping constant the space coordinates (x^1, x^2, x^3) . In this system the trajectories of a^i are the self axis x^0 , and $a^1 = a^2 = a^3 = 0$. On choosing a unit Killing vector,

$$a^i(x) \equiv (1,0,0,0) \,,$$

we shall find Killing's equation reduced to:

$$a^i \frac{\partial g_{jk}}{\partial x^i} = a^0 \frac{\partial g_{jk}}{\partial x^0} = 0 \,, \text{ or,}$$

$$\boxed{\frac{\partial g_{jk}}{\partial x^0} = 0}$$

In the new coordinate system all the components of the metric tensor are time independent. We say that the coordinate system is adapted to the stationary character of the metric, i.e., it is adapted to the Killing vector field. Kerr's metric which describes a rotating body is an example of stationary metric.

Static Field

A given gravitation field is called static, if there exists a coordinate system adapted to the stationary character of the metric for which

$$g_{0\nu} = 0 \ \ (\nu = 1, 2, 3 \ ) \text{ and}$$

$$\frac{\partial g_{ij}}{\partial x^0} = 0 \ .$$

The geometric interpretation is that the trajectory of the Killing vector fields $a^i(x)$ are, for a static field, orthogonal to a family of hypersurfaces.

Mathematical Note.

Given an equation of the type:

$$F(x^1, ..., x^N) = 0,$$

we say that it determines an hypersurface in N-dimensional space. In fact, by the rules of multi-variable Calculus:

$$dF = 0 \ \ \therefore \ \ \frac{\partial F}{\partial x^i} dx^i = 0.$$

In other words, for each displacement on the surface, the gradient vector $\frac{\partial F}{\partial x^i}$ which is covariant, is orthogonal to the hypersurface. In that way, the Killing vector $a^i(x)$ will be orthogonal to a given family of hypersurfaces $\sigma(x) = 0$ if it can be expressed by a relation of the type:

$$a_i(x) = \chi(x) \frac{\partial \sigma(x)}{\partial x^i} \ ,$$

where $\sigma(x)$ and $\chi(x)$ are scalar functions. Schwarzschild's metric is static according with this definition.

Part III

Introduction to Relativistic Cosmology

Chapter 6

Digression into Philosophical and Mathematical Matters (Optional Study)

6.1. Newtonian Concepts of Space and Time

Newton professed a philosophical conception over space and time, originated from the theories of Henry More (Koyré, 1962). Notwithstanding Newton having looked how to justify his space conception in the context of Physics, with the critical analysis of the notion of inertial force, he employed Henry More's philosophical system as a supplemental argument which seems made in order to reinforce his scientific conception. To Henry More, space and time should be absolute and infinite because they were manifestation of a "divine entity". This does not seem to us a serious argument because it mixes the quantitative infinity with the ontological infinity and because it materializes time and space along with the "divinity". More than that, as we shall show later, the philosophical analysis of gravitational and inertial phenomena shows that spacetime is a form connected with matter. This analysis, which encompasses Mach's Principle, was the departure point from which Einstein developed his covariant theory of Gravitation, which is known as the GRT (General Relativity Theory).

A basic principle of Newtonian dynamics, and also of the Special Theory of Relativity, is the relativity of the inertial systems of references. This means: there are no privileged reference systems, or, in other words, each one is equivalent to any other. We can say that the dynamical laws in those theories are invariant under inertial reference systems. This principle does not apply to accelerated reference systems. Therefore, Newton's second law, which says that force is the temporal derivative of linear momentum, is not anymore universally valid. This means that for coordinate transformations involving accelerated observers, there is a surge of other kinds of force, the so called inertial forces. The universal character of dynamical laws is now broken. More than that, because Newton's second law represents the principle of efficient causality ("give me the time derivative of linear momentum and I will show you the force"), now there is a problem as to the validity of this principle which was pointed by Einstein (1923): why should exist forces responsible for mere coordinate transformations and which do not respond for any known interaction?

Let us examine what happens with the rotation motion of a particle. Consider two observers $\sum$ and $\sum'$ one in rotation relative to the other. If the relativity principle would be valid, each one of both observers could tell to the other one: "I am at rest, and you are in rotational motion". This is not verifiable in Newtonian physics because one of those observers faces acceleration, while the other one does not. Consequently, Newton, in order to save the relativity principle, postulated the existence of an absolute observer so that he could distinguish $\sum$ from $\sum'$. The presence of an acceleration in $\sum'$ and not in $\sum$, is due to the existence of an absolute space, relative to which a given referential can either be accelerated or not. But as Einstein (1923) observed, this does not put the principle of causality in a safe ground. We conclude that, within the Newtonian theory of absolute space, the principle of causality becomes mutilated. The reader should remember that Einstein substituted, in Special Relativity, absolute time, and absolute space, for an absolute spacetime, represented by an invariant line element (Minkowski's). In General Relativity Theory, this "absolutism", was reduced to a local concept, in a Riemannian space.

In the next section we shall follow arguments against a theory of cosmology made up from Newtonian gravitational theory.

6.2. Objections to Newton's Cosmology

The basis for Newtonian Cosmology is Poisson's equation for the gravitational field. For readers not familiar with this equation we refer to basic university physics texts, where, in electrostatics, there is an analogous equation:

$$\nabla^2 U = 4\pi G\rho \ . \qquad (6.2.1)$$

The above equation for the Universe, presupposes that matter is continuously distributed with mass density ρ , while G stands for Newton's gravitational constant and U is the gravitational potential. We justify the matter hypothesis above, because the mass distribution is extended over cosmic distances.

Einstein showed that this equation does not substitute perfectly the action-at-a-distance gravitational law because of the boundary conditions for the solution of the equation. In other words, we have to impose boundary conditions for the gravitational potential U at infinity. The solution of Poisson's equation requires that, when $r \longrightarrow \infty$, we should have $\lim_{r\to\infty} U = U_\infty =$ fixed value. We conclude that $\lim_{r\to\infty} \rho = 0$.

Nevertheless, we know that $U \sim r^{-1}$ (Halliday, Resnick and Walker, 2005). So $\rho \sim r^{-3}$. This means that ρ goes to zero more rapidly than U, for large values of r. This behavior led Einstein to say that Newtonian cosmology models possesses infinite mass, but in some sense it is finite because the gravitational potential goes to zero more slowly than the density!!! Let us now examine why Poisson's equation was not acceptable to Einstein in the cosmological domain.

First: the inhomogeneity due to $\rho \sim r^{-3}$ is incompatible with astronomical observations.

Second: it includes the notion of "center of the Universe" which is meaningless.

Third: Einstein shows that $\rho \sim r^{-3}$ is against statistical physics theory. The boundary condition imposes - as we saw – to Poisson's equation, that $\lim_{r \to \infty} U = U_\infty =$ fixed value. Now, by statistical physics, each astronomical object could acquire kinetic energy enough to overcome this value and then disappear towards infinity. Thus we would not find $\rho \sim r^{-3}$. We see therefore, that the static character of this Universe which is represented by Poisson's equation, is inviable.

Fourth: We could find an alternative solution $\rho =$ constant, but this can not be sustained by Poisson's equation because the gravitational potential U would become undetermined. This objection was presented by Seeliger in the year 1895, as cited in North (1965).

Fifth: In order to make superseded, the Seeliger's objection, it could be suggested that there is a universal constant λ , in a modified Poisson's equation, namely:

$$\nabla^2 U - \lambda U = 4\pi G\rho \ . \tag{6.2.2}$$

A possible solution to this equation would be:

$$U = -\frac{4\pi G}{\lambda}\rho \ , \tag{6.2.3}$$

with $\rho =$ constant.

However this modification introduces a new constant λ by means of an ad-hoc hypothesis. This runs against the theoretical framework.

Sixth: As Einstein observed, the radiating energy is originated by mass and it gets lost into infinity, which is also incompatible with the hypothesis $\rho \sim r^{-3}$.

Seventh: The functions $U \sim r^{-1}$, and $\rho \sim r^{-3}$, carry a nonsense: at infinity we would have a constant value U_∞ and the absence of mass because ρ goes to zero faster than U goes to U_∞. Einstein interpreted this hypothesis as being a universe with infinite mass, contained in a finite volume. Obviously, the mathematics is inconsistent: a universe in which $\lim_{r\to\infty} U = U_\infty =$ constant, with $\rho_\infty = 0$, while $U \sim r^{-1}$ and $\rho \sim r^{-3}$, gives us the idea that there is an exterior space outside the universe, because in physical theory we can not speak rigorously about infinities. When we say that $\lim_{r\to\infty} \rho = 0$, we simply represent mathematically the idea that for large distances r , the density goes to zero. Therefore such infinite limit brings about the concept of exterior space around the Universe, whose intrinsic antinomy was pointed by Nicolaus Cusa (1954), Copernicus, Descartes (Koyré, 1962) and others.

The idea of exterior space outside the Universe implies the concept of a Universe limited by a surface separating matter from vacuum, which is a senseless proposition. In fact, space as such, is a set of all possible positions of bodies relative to each others, i.e., a common form of phenomena (Weyl, 1950). As Saint Augustine (1958) remarked, the place of the universe is in itself , and there is no sense in talking about the Cosmos' exterior.

6.3. Mach's Principle

We saw, in the last two sections, the objections that can be raised against Newton's conception of infinite and absolute space, as well as the contradictory character of a cosmology based in his gravitational theory. Ernst Mach (1912), at the end of the nineteenth century, saw the difficulty involved in the justification given by Newton to the existence of an absolute reference system independent of the matter field, based on the consideration of local inertial forces. Logically, Mach refuted Newton's conclusion, on grounds that a reference system can only have physical significance, if it is tied to a matter field. He concluded, therefore, that local inertial accelerations should be referred to an inertial cosmic frame, defined by the distant masses of the Universe.

Einstein (1923), considering Mach's criticism, and, taking the causality principle in the definition of forces into account, inferred that local inertial forces are determined by a gravitational interaction between the local system and the distribution of cosmic masses. This is the famous Mach Principle, as coined by Einstein. So, Mach Principle encompasses two fundamental elements, as cited by Brans and Dicke (1961) and Gomide (1973):

1) Local accelerations are related to an inertial cosmic reference system determined by the distribution of the (distant) masses of the Universe.

2) These accelerations, which are responsible for inertial forces, are the result of a gravitational interaction between the local system and the (distant) cosmic masses. Inertial effects are connected with the gravitational interaction.

Mach Principle as formulated by Einstein, requires that the content of the causality principle for the existence of forces can not be justified by a mere appeal to an absolute reference frame, because these forces have their *raison d' être* in the reality of a physical interaction. The explanation given by Newton, as we saw earlier, hurts a fundamental principle of ontology. Einstein's philosophical view is much more penetrating than that of his XVIIth century colleague. Robert Dicke (1967) observes that Einstein succeeded better, when he based his GRT in dense philosophical analyses.

Mach Principle implies that a cosmic reference frame is determined by the distribution of masses in the Universe. Now, relativistic physics shows us that the material processes occur in a four dimensional continuum. By consequence, time must be associated to any frame in the universe. This fundamental notion may induce a more penetrating view of the problem caused by the existence of inertial forces in local reference frames, and to the

better understanding of the difference between of the proper time and coordinate time, as referred to Mach Principle.

Given one local reference frame not subject to accelerations we could employ cylindrical coordinates in order to define the usual metric of Special Relativity. Let it be:

$$d\tilde{s}^2 = -\left[d\tilde{r}^2 + \tilde{r}^2 d\tilde{\phi}^2 + d\tilde{z}^2\right] + c^2 dt^2 . \qquad (6.3.1)$$

Consider now a rotating frame, such that:

$$\begin{aligned} z &= \tilde{z} \\ r &= \tilde{r} \\ t &= \tilde{t} \\ \phi &= \tilde{\phi} - \omega t \quad . \end{aligned}$$

Let ω be the angular velocity, so that the new metric is given by:

$$ds^2 = -\left[dr^2 + r^2 d\phi^2 + dz^2\right] + 2r^2\omega d\phi dt + (c^2 - \omega^2 r^2) dt^2 . \qquad (6.3.2)$$

Consider now two observers at rest in both reference frames, respectively; we shall have:

$$d\tilde{s}^2 = c^2 dt^2 . \qquad (6.3.3)$$

$$ds^2 = (c^2 - \omega^2 r^2) dt^2 . \qquad (6.3.4)$$

We see that the proper times in both frames are different from one another, and related by:

$$ds^2 = (c^2 - \omega^2 r^2)(d\tilde{s}^2 / c^2) . \qquad (6.3.4.a)$$

We may infer that the problem which left Newton perplex, i.e., the impossibility of application of the relativity principle from inertial to rotating frames is originated by a property ignored at the time, namely that time is relativistic and not absolute. The duality between proper time and coordinate time is underlined in the grounds of rotating frames and it can be perceived by the privileged character of the frame under centrifugal acceleration; its consequence is the existence of a metric component $g_{00} \neq 1$, and given by:

$$g_{00} = (c^2 - \omega^2 r^2)/c^2 \quad , \tag{6.3.5}$$

which entails a local inertial effect. The accelerating reference frame possesses a proper time different than that of the non accelerating system. Now, Mach Principle, as seen from the relativistic point of view, must also mean the determination by the cosmic masses, of a common universal time with the local observer (Berman, Gomide, Garcia, 1986). Once that the time measurement is not independent of matter, the cosmic mass should be in a global motion (like, for instance, expansion) capable of defining such time. It must be observed that a static cosmic distribution of masses could not define a "time": for a "time" in a static universe would not be different than the Newtonian absolute time independent of matter, i.e., it could not be defined because, as we pointed above, time is tied to motion. What we saw, namely:

$$ds^2 = g_{00}(r) d\tilde{s}^2 \ , \tag{6.3.6}$$

it is there to signify an interaction of the local rotating system with the cosmic system in motion, via the liaison between both proper times. Consequently the metric component $g_{00}(r)$ which represents an inertial effect is also significant of an interaction of the local accelerated system with the distribution of masses in the Universe. From this original visualization of Mach Principle, according to the relativity of time, we must conclude that for the principle to be realized, the cosmic masses must be in motion, unless we fall into the untenable idea of an absolute Newtonian cosmic time. The interaction property inherent to $g_{00}(r)$ attaches a special significance to the gravitational interaction potential, when we think of the equivalence principle (see section 6.5).

In the end of this Chapter,it will be shown, as we shall also point out in later Chapters, that a mathematical formulation of Mach's Principle may be assumed to be the statement, (and its mathematical consequences) that the total energy of the Universe is zero-valued. (Berman, 2006; 2006 a).

D'Inverno summarizes Machian ideas with three statements (D'Inverno, 1993):

I - the matter distribution determines geometry.
II - "no matter" means "no geometry".
III - in an empty Universe, a body would have no inertia.

6.4. The Potential for Gravitation

From Mach Principle, which denies the existence of Newtonian absolute space, and them postulates the equivalence of inertia and gravitation in the cosmic domain, it is only a question of mere necessity that a new formulation be made for the gravitational theory. We have seen that Newton's scalar potential presupposes the idea of an absolute space dissociated from matter, and without agreeing with any equivalence between inertia and gravitation. Dicke (1964) studied some possibilities for a possible gravitational potential that would obey Mach Principle, while showing that the best candidate is a tensorial potential associated with the metric tensor. Let us follow the line of thought established by Dicke:

First option: **a scalar potential in a quadridimensional spacetime.**

In this case a quadriforce acting on a particle due to a scalar field would be given by:

$$F_\alpha = -\phi_{,\alpha} \,. \qquad (6.4.1)$$

Newton's second law should be:

$$F_\alpha = \frac{dp_\alpha}{d\tau} \,, \qquad (6.4.2)$$

where p_α is the linear quadrimomentum.

By hypothesis, the scalar potential is not time-dependent so that:

$$\phi_{,0} \equiv \frac{\partial \phi}{\partial x^0} = 0. \qquad (6.4.3)$$

Now, this means that the energy of the particle, i.e., the time component of the quadrimomentum, is:

$$p_0 = m(1-\beta^2)^{-1/2}c^2 = \text{constant}. \qquad (6.4.4)$$

Now, the existence of a force acting on the particle means that the kinetic energy has to vary with time, which is contradictory to p_0 being a constant.

More than that, in order for p_0 to be constant it would be necessary that its rest mass should vary with time which contradicts the concept of the rest mass of a particle in Special Relativity.

Second option: **Vector potential in quadri-dimensional spacetime.**

This would be a potential in an analogy with that of the electromagnetic field. Such a vector potential would determine a privileged direction in space because the rotation part of the field carries field lines of a non isotropical paradigm. (The reader can easily think that there is one privileged direction in spacetime: that of the vector potential). Now, the local inertial effects posses an isotropic character. [Weinberg (1972) discusses nevertheless, the study of anisotropy of inertia].

Third option: **Tensor potential.**

In Classical Mechanics the equations of motion can be derived from Hamilton's Principle (Goldstein, 1980):

$$\delta \int [K - V]\, dt = \delta \int L dt = 0. \qquad (6.4.5)$$

where K, V and L are the kinetic energy, potential energy and the Lagrangian of the problem. We would be tempted to write for the motion of a particle in the gravitational field an equation of the type:

$$\delta \int [I + G]\, d\tau = 0\,, \qquad (6.4.6)$$

where $L = I + G$. (We separated the Lagrangian into two parts: the inertial I and the gravitational G). However, from Mach Principle, due to the equivalence of inertia and gravitation, we may convert I into G and vice-versa. So we would write:

$$\delta \int G d\tau = 0 \qquad (6.4.7.a)$$

$$\delta \int I d\tau = 0 \quad . \qquad (6.4.7.b)$$

Then we must look for a Lagrangian quadratic in velocity which represents the kinetic part $K$of the motion, but whose terms are multiplied by terms representative of the potential for gravitation or the inertia potential.

So, we must drop the relation $L = K - V$ with preference to a relation of the type $L = KV$. This product evidences the connection of the inertial effects of kinetic origin, with gravitation or inertia. Gravitation and inertia appear in a mixed form. Then this Lagrangian must assume the form:

$$L = g_{\alpha\beta} u^{\alpha} u^{\beta}\,. \qquad (6.4.8)$$

This means that the gravitational or inertial potential is a second order tensor $g_{\alpha\beta}$. The most simple, (but representative of Mach Principle), variational principle would be:

$$\delta \int g_{\alpha\beta} u^{\alpha} u^{\beta} d\tau = 0. \qquad (6.4.9)$$

Notice that we use proper time as the integration variable. This principle is formally equivalent to the equation of geodesics in Riemannian space:

$$\delta \int ds = 0 \,, \tag{6.4.10}$$

from which, as it is well known, we derive the metric form of the geodesics,

$$ds^2 = g_{\mu\nu} dx^\mu dx^\nu \,, \tag{6.4.11}$$

which can also be written:

$$g_{\mu\nu} \frac{dx^\mu}{ds} \frac{dx^\nu}{ds} = 1 \,. \tag{6.4.11.a}$$

We remember that the quadrivelocities are defined by:

$$u^\alpha = \frac{dx^\alpha}{d\tau} \,.$$

Mach Principle, therefore, suggests that the tensor potential for gravitation is equivalent to the Riemann metric tensor. So, gravitation and inertia determine the geometry of spacetime.

6.5. Equivalence Principle

Another principle which connects gravitation, inertia, and spacetime geometry, is the equivalence principle, which has an experimental basis. Let us call $\vec{g}$ – the acceleration of gravity; and m_i the inertial mass of a particle, i.e., the mass that appears in Newton's second law. If m_g stands for the gravitational mass, which is the mass that appears in the acceleration of gravity or with Newton's Law of Gravitation, we may write:

$$m_g \vec{g} = m_i \vec{a} \,, \tag{6.5.1}$$

where $\vec{a}$ stands for the acceleration of the particle subject to the gravitational field. From the above we write:

$$\vec{a} = (m_g / m_i) \vec{g} \,. \tag{6.5.2}$$

It would seem that the acceleration $\vec{a}$ would be different for the different bodies with different values of m_g/m_i. Galileo in the sixteenth century verified, with the precision limits available at his time, that the acceleration $\vec{a}$ for bodies in free fall independs on the masses of those bodies. Newton, in the next century, tried pendulums of the same length, same mass and different materials, and did not find any difference in their respective periods. Friedrich Wilhelm Bessel, one century later, confirmed Newton's results with a better precision. In the last century, Roland von Eötvös (Eötvös et al, 1922) verified that the ratio m_g/m_i independs on the nature of those bodies, with a precision of 5 in 10^9. The experiment was repeated by Dicke (1964) with the precision 1 in 10^{11}, with the same result: the mentioned ratio independs on the nature of the bodies. Though all these experiments only prove that gravitational and inertial masses are proportional to each other, without loss of generality, we can set the proportionality as an equality, and write:

$$m_g / m_i = 1 \ . \tag{6.5.3}$$

We can conclude that the path followed by a body subject to inertia and gravitation, independs on the formal atomic structure of the body. In this regard, inertia and gravitation are only related with the trajectory of a body in spacetime. Now, spacetime is the common ground for relationships among physical phenomena (Weyl, 1950), which means that inertia and gravitation are only connected with physical geometry. Saying that otherwise, the geometry of spacetime is the common form of the matter in the interaction of inertia with gravitation, which is, the universal property of matter; their properties can be expressed in terms of the geometry. To say that the path of a body is solely determined by gravitation and inertia, is the same thing as to say that this path is solely determined by the geometry. Which curves are those which only depend on spacetime? – These are the geodesics. In Euclidean geometry the geodesics are straight lines. But a body like a planet subject to gravitation and inertia, does not move in a straight line. This means that the geodesics must be a family of curves in non-Euclidean geometry. The obvious choice is Riemannian spacetime, whose geodesics are given by:

$$\delta \int ds = \delta \int \sqrt{g_{\alpha\beta} u^\alpha u^\beta} ds = 0 \ . \tag{6.5.4}$$

This results in:

$$\tfrac{d}{ds}(g_{\alpha\beta} u^\beta) - \tfrac{1}{2} g_{\beta\delta,\alpha} u^\beta u^\delta = 0 \ ,$$

while

$$ds^2 = g_{\alpha\beta} dx^\alpha dx^\beta \ .$$

We may write, taking into account that $m_g = m_i$,

$$m_i \tfrac{du_\alpha}{ds} = \tfrac{1}{2} m_g g_{\beta\delta,\alpha} u^\beta u^\delta \ . \tag{6.5.5}$$

In this last form, we have a generalized Newton's second law where the inertial force is equivalent to a force arising from a gradient of the tensorial potential $g_{\alpha\beta}$. Now, the metric coefficient of the geometry has the physical meaning of a tensor potential. We have found the law of free fall in a uniform gravitation field, from the equations of geodesics.

D'Inverno (1993) states that the above implies also in five statements :

I - the motion of a small body in a gravitational field independs on its mass and composition.

II - matter is acted by gravitational fields, but is also a source of such fields.

III - the gravitational field is coupled to everything else because it contains energy.

IV - the concept of inertial frame is localized.

V - a rest frame immersed in a gravitational field, is equivalent to a reference frame linearly accelerated in Special Relativity.

6.6. The principles of (General) Relativity, Covariance, and Simplicity

As Mach Principle denies the existence of privileged systems of reference, all observers must be equivalent, and not only the inertial ones. There should be a *principle of general relativity* for arbitrary transformations of coordinates. We are led to choosing the line element of Riemann geometry, namely:

$$ds^2 = g_{\mu\nu} dx^\mu dx^\nu \ ,$$

because this is an invariant quantity for arbitrary coordinates.

We have seen earlier, that the metric tensor has to do with accelerations. This results from Mach and Equivalence Principles. These two principles make us consider the metric tensor of Riemann geometry as applicable to Physics in order to represent the gravitational field or the accelerations field. The coefficients of the metric tensor assume physical properties. This shows that Newtonian theory, which considers physical spacetime as an empty and absolute spacetime which has nothing to do with the matter fields, should be wrong. As in the Greek Philosophy of Aristotle and Plato space and time are seen as a category of matter.

The **principle of covariance**, for physical laws, may be thought as devoid of physical contents, for one could take, a given relation, valid only in a particular system, and write it in tensor form, but if the physics involved would not allow for a correct translation into a covariant phenomenon, the tensor equation would be devoid of validity. Nevertheless, correct physical laws must be written in tensor notation, which is the best way to know how to express it in other coordinate systems, with equal validity.

We have seen that, in Special Relativity, the energy momentum tensor obeys a conservation law:

$$T^{\mu\nu}_{,\nu} = 0 \tag{6.6.1}$$

In General Relativity, the obvious generalization would be:

$$T^{\mu\nu}_{;\nu} = 0 \tag{6.6.2}$$

However, there remain other possibilities, involving the curvature tensor, which is null in Special Relativity; for instance (D'Inverno, 1993):

$$T^{\alpha\beta}_{;\beta} + g^{\beta\varepsilon} R^{\alpha}_{\beta\gamma\delta} T^{\gamma\delta}_{;\varepsilon} = 0 \tag{6.6.3}$$

We point out that simplicity requirements imply in "minimal coupling" , so that we prefer (6.6.2) as the correct generalization of (6.6.1): no terms including the curvature tensor should, then, be added in the generalization.

By the same token, we do not admit, in General Relativity, variations, either in time or in space, of the constants in the theory:

$$\frac{\partial G}{\partial x^{\alpha}} = \frac{\partial m_0}{\partial x^{\alpha}} = \frac{\partial c}{\partial x^{\alpha}} = \frac{\partial \varepsilon_0}{\partial x^{\alpha}} = \frac{\partial \mu_0}{\partial x^{\alpha}} = 0 . \qquad (\alpha = 0,1,2,3)$$

In the above, G, m_0 , c , ε_0 , μ_0 are respectively Newton's gravitational constant, rest mass, speed of light, permittivities of electric and magnetic origin (the reader should be acquainted with the usual denomination of the latter, *permeability constant*; I prefer to call it magnetic permittivity, but this denomination, is my own).

Of course, removing the first term requirement makes for Brans-Dicke theory; the second one, for Kaluza-Klein-Wesson penta-dimensional theory; the other three, for Albrecht-Magueijo's. (Brans and Dicke, 1961; Wesson, 1999, 2006; Albrecht and Magueijo, 1998).

The requirement of simplicity, does not mean that we should ignore experimental observations; the cosmological constant, was dismissed from the theory some time ago, on this requirement: however, cosmological observations asked recently for its reinstatement. A variable G or Λ , may be necessary from physical requirements. An affinely connected space is described by torsion and two curvatures, the old Riemannian , and the homothetic (or segmentary), in accordance with Élie Cartan´s book on affine connections (Cartan, 1986).

Riemannian geometry might eventually have to be replaced by this new geometry which included torsion (Berman and Marinho, 1996). The simplicity requirement could only be used to keep the first geometry, if there were no physical evidence to the contrary; again, the simplicity requirement could be used, for instance, to choose between the tridimensional possibilities for the physical world, and express that it is flat; or to decide between isotropic or anisotropic models of the Universe, in favor of the former; or between homogeneous or inhomogeneous models of the Universe, choosing the former, etc.

6.7. Principles of "cosmic time" and "correspondence"

For consistency, we must impose a correspondence law, contained in the statement:

"General Relativity reduces to Special Relativity in the absence of gravitation; it reduces to Newtonian gravitation theory in the case of weak gravitational field and low speeds; and to Newtonian mechanics, in the absence of gravitation, and for low speeds".

Consider the metric line-element:

$$ds^2 = g_{\mu\nu} dx^{\mu} dx^{\nu} \qquad (6.7.1)$$

If the observer is at rest,

$$dx^i = 0 \qquad (\ i = 1,2,3 \ ) ,$$

while,

$$dx^0 = dt \ . \tag{6.7.2}$$

This last equality defines a proper time; we called cosmic time, in Cosmology.

From the geodesics' equations, we shall have:

$$\frac{d^2x^i}{ds^2} + \Gamma^i_{\alpha\beta}\frac{dx^\alpha}{ds}\frac{dx^\beta}{ds} = \Gamma^i_{00} \ . \tag{6.7.3}$$

We then find:

$$g^{ij}\frac{\partial g_{i0}}{\partial t} = 0 \ . \tag{6.7.4}$$

This defines a Gaussian coordinate system, which in general implies that:

$$\frac{\partial g_{i0}}{\partial t} = 0 \ . \tag{6.7.5}$$

We must now reset our clocks, so that, the above condition is universal (valid for all the particles in the Universe), and then our metric will assume the form:

$$ds^2 = dt^2 - g_{ij}\left(\vec{x},t\right) dx^i dx^j . \tag{6.7.6}$$

If we further impose that, in the origin of time, we have:

$$g_{i0}(t = 0) = 0 \ , \tag{6.7.7}$$

then by (6.7.5), we shall have:

$$g_{i0}(t) = 0 \ . \tag{6.7.8}$$

The above defines a Gaussian normal coordinate system.

For a comoving observer, in a freely falling perfect fluid, the quadrivelocity u^μ will obey:

$$u^i = 0 \ , \tag{6.7.9}$$

while, if we normalize the quadrivelocity, we find, from the condition:

$$g_{\mu\nu}u^\mu u^\nu = 1 \ , \tag{6.7.10}$$

that,

$$g_{00}u^0 = 1 \ . \tag{6.7.11}$$

Though Gomide and Berman (1988) have discussed the case $g_{00} = g_{00}(t) \neq 1$, we usually impose:

$$g_{00} = u^0 = 1 \ . \tag{6.7.12}$$

The purpose we have in mind, is to define a Machian metric; Gaussian coordinate systems, in fact, imply that, with $g_{0i} = 0$, there are no rotations in the metric (we refer to formula (6.3.3) to the contrary), and in each point we may define a locally inertial reference system.

Gaussian normal coordinates were called "synchronous" ; in an arbitrary spacetime, when we pick a spacelike hypersurface S_0, and we eject geodesic lines orthogonal to it,

with constant coordinates x^1, x^2 and x^3, while $x^0 \equiv t + t_0$, where $t_0 = 0$ on S_0, then t is the proper time, whose origin is $t = 0$ on S_0 (see MTW, 1973).

In the above treatment, cosmic time is "absolute", so that the measure of the age of the Universe, according to this "time", is not subject to a relative nature.

As it was remarked in Chapter 3, a GRT formula involving covariant derivatives, reduces, in its Special Relativistic equivalent, to partial derivatives. The converse, however, may not be true. Also, to each point in a Riemannian space, there corresponds a "tangent" plane, where Special Relativity laws, are locally valid. This is not a global property. Geodesic coordinates, are meant to reproduce a nearly flat-space frame around a given point. If a tensor equation, is shown to be valid in such coordinates, it is automatically valid for any reference system, where the mathematics is more complex. We remember that, for geodesic coordinates, defined in a given point, the Christoffel symbols are locally null, at the same point. The equivalence principle, then, means that locally, we can not decide whether we have a non-rotating free-fall state, in the presence of a gravitational field, or uniform motion in its absence. A linear accelerated frame, relative to an inertial one, in the absence of gravitation, is locally indistinguishable from a rest-frame, immersed in a gravitational field.

Final Observations:

1st) Not all metric elements from Riemannian Geometry are adapted for representing the gravitational field of GRT; in fact we must have a "spacetime" metric, which is defined in order that, locally, it should reduce to Minkowskian's metric. Other arbitrary 4-dimensional Riemannian spaces can not represent spacetime, because of the requirement that GRT should reduce to Special Relativity.

2nd) The difference between GRT's spacetime fit to represent a gravitational field, and the one of Special Relativity, is that, in at least one infinitesimal neighborhood of a spacetime "event", the Ricci and Riemann tensors must be non-null.

3rd) In GRT, the energy-momentum tensor $T_{\mu\nu}$ must at least have one non-zero component in at least one infinitesimal neighborhood of a spacetime "event" or else, we shall be in a zero-gravitational case.

4th) The energy-momentum tensor employed in GRT must reduce to a symmetric tensor in Special Relativity.

6.8. Revisiting Mach's Principle

Consider now, that we write the energy equation for a particle of mass m, subject to the gravitational mass M, of the "spherical" Universe,of observable radius R.The inertial energy, for the sum of all such particles, is given by,

$$E_i = \sum mc^2 = Mc^2,$$

while the sum of the potential energies,

$$E_{pot} = -\sum \frac{GMm}{R} = -\frac{GM^2}{R}$$

If we consider that inertia "here" is gravitation "there", we may think that the two energies cancel each other,and the total is, for the whole Universe, zero-valued. We obtain the Brans-Dicke relation,

$$\frac{GM}{c^2R} = \gamma \approx 1$$

Chapter 7

Introductory Cosmological Models

7.1. Derivation of the Homogeneous and Isotropic Metric

We want to obtain a model of the Universe, capable of making possible a comparison between theoretical parameters and astronomical experimental data. It is necessary to use the argument of simplicity , because on observing the Universe as a whole, we find many "small" complexities, so that the big picture can only be obtained under simplifying hypotheses. The first one, which we now present, is known as the Cosmological Principle:

There are no privileged observers in the Universe.
All points and all directions are equivalent.

In other words, the Universe must be represented by a metric which, in its tridimensional space contents, is homogeneous and isotropic.

From the "simplicity" criterion stems the perfect fluid hypothesis, namely, that the matter that is contained in the Universe can be represented by the perfect fluid energy tensor. For a comoving observer, that, at a given cosmic time "t", is "at rest" in the given coordinate system:

$$T^{ij} = p\delta^{ij} \qquad ; \qquad (\, i,j = 1,2,3)$$

$$T^{i0} = T^{0i} = 0\,;$$

$$T^{00} = \rho\,,$$

where ρ and p stand for the energy density and cosmic pressure respectively. Hence, the comoving observer sees the fluid around him as isotropic, and the reference system is locally inertial. It has been verified (see section 7.4) that the Universe is not static, but in fact, it is expanding; in Special Relativity, the metric employed was Minkowski's. We now write it in a slightly different form, by including a scale factor $R(t)$ so that the differentials dx, dy, dz suffer a dilatation with increasing time:

$$ds^2 = dt^2 - R^2(t)\left[dx^2 + dy^2 + dz^2\right]. \qquad (\, 7.1.1\,)$$

As we know well, Minkowski's metric is a flat metric; the above metric has spatial part:

$$d\sigma^2 = R^2(t)[dx^2 + dy^2 + dz^2] \,, \qquad (7.1.2)$$

which is obviously also flat. In fact, this is a zero curvature submetric. Metric (7.1.1) is called Robertson-Walker's metric for a flat Universe.

For a three-dimensional space with curvature, the simplest case is that of "constant curvature". Its Riemann-Christoffel tensor is given by:

$$R_{abcd} = k(g_{ac}g_{bd} - g_{ad}g_{bc}) \,, \qquad (7.1.3)$$

where k stands for the tri-curvature. For the flat metric $k = 0$. (the reader can check this easily). Its Ricci tensor is then given by:

$$R_{\beta\delta} = g^{\alpha\gamma}R_{\alpha\beta\gamma\delta} = 2kg_{\beta\delta} \,. \qquad (7.1.4)$$

In tridimensional space, like in the derivation we made for the spherically symmetric Schwarzschild's metric, we now expect, a spatial part of the metric in the form:

$$d\sigma^2 = e^{\alpha(r)}dr^2 + r^2(d\theta^2 + \sin^2\theta d\phi^2) \,. \qquad (7.1.5)$$

For this metric, we find:

$$R_{11} = \frac{\alpha'}{r} \,; \qquad (7.1.6)$$
$$R_{22} = \text{cosec}^2\theta \, R_{33} = 1 + \tfrac{1}{2}re^{-\alpha}\alpha' - e^{-\alpha} \,.$$

Then, the condition of constant space curvature reduces to the following equations:

$$\frac{\alpha'}{r} = 2ke^{\alpha}, \qquad (7.1.7)$$

$$1 + \tfrac{1}{2}re^{-\alpha}\alpha' - e^{-\alpha} = 2kr^2 \quad . \qquad (7.1.8)$$

The solution for the above is:

$$e^{-\alpha} = 1 - kr^2 \,. \qquad (7.1.9)$$

The expanding spatial part would now become:

$$d\sigma^2 = R^2(t)[\tfrac{dr^2}{1-kr^2} + r^2(d\theta^2 + \sin^2\theta d\phi^2)] \quad . \qquad (7.1.9a)$$

We now compare with the metric (7.1.2), which in spherical coordinates would have the form:

$$d\sigma^2 = R^2(t)[dr^2 + r^2(d\theta^2 + \sin^2\theta d\phi^2)] \,. \qquad (7.1.10)$$

We said the this last one was a flat metric, which now is shown to belong to the case $k = 0$ (zero tricurvature). We are thus led to the Robertson-Walker's metric for the Universe,

$$ds^2 = dt^2 - R^2(t)[\frac{dr^2}{1-kr^2} + r^2(d\theta^2 + \sin^2\theta d\phi^2)] \ . \qquad (7.1.11)$$

For this metric, the 3-dimensional curvature scalar is given by:

$$^3K(t) = kR^{-2}(t) \ . \qquad (7.1.12)$$

Basically, we have three possible cases in Cosmology: $k = 0$; $k > 0$; and $k < 0$, representing flat, positively curved or negatively curved spaces.

For practical reasons one can express Robertson-Walker's metric in a slightly different form:

$$ds^2 = dt^2 - \frac{R^2(t)}{[1+\frac{k\bar{r}^2}{4}]^2}[d\bar{r}^2 + \bar{r}^2(d\theta^2 + \sin^2\theta d\phi^2)] \ , \qquad (7.1.13)$$

where we made use of the substitution:

$$r = \frac{\bar{r}}{1+\frac{k}{4}\bar{r}^2} \ . \qquad (7.1.14)$$

Usually, the bar in the "new" radial coordinate are dropped from the expression of the metric, and the new 3-curvature k' is also introduced, by means of a re-scaling:

$$k = |k|k' \ , \qquad (7.1.15)$$

so that, the three cases above are expressed by $k' = 0, +1$, or -1 . We then rescale the radial coordinate, by:

$$\bar{\bar{r}} = \sqrt{|k|}\bar{r} \ , \qquad (7.1.16)$$

and also we rescale the scale factor, by:

$$\bar{R}(t) = \frac{R(t)}{\sqrt{k}} \quad \text{for } k \neq 0 \quad ,$$

or, (7.1.17)

$$\bar{R}(t) = R(t) \text{ for } k = 0 \quad .$$

We still keep the same form of the metric, on dropping bars again, so that Robertson-Walker's metric can either be expressed in each of these alternative forms:

$$ds^2 = dt^2 - R^2(t)[\frac{dr^2}{1-kr^2} + r^2(d\theta^2 + \sin^2\theta \ d\phi^2)] \quad , \qquad (7.1.15)$$

or

$$ds^2 = dt^2 - \frac{R^2(t)}{[1+\frac{k}{4}r^2]^2}[dr^2 + r^2(d\theta^2 + \sin^2\theta \ d\phi^2)] \quad . \qquad (7.1.16)$$

The cosmological models which follow from this metric, in the case of null Cosmological Constant ($\Lambda = 0$), are called FRW's models (Friedmann - Robertson - Walker models); in the case of non zero Cosmological Constant ($\Lambda \neq 0$) they are called Lemaître's models.

7.2. Einstein's Field Equations for Cosmology

In the year 1922, the Russian cosmologist Friedmann, and then in 1927, the Belgian Abbé Lemaître, applied Einstein's field equations into the above metric.

Consider R.W's metric given by (7.1.15). On calculating the Christoffel symbols, we find the non-vanishing terms:

$$\Gamma^0_{11} = \frac{R\dot{R}}{1-kr^2}$$

$$\Gamma^0_{22} = R\dot{R}r^2$$

$$\Gamma^0_{33} = R\dot{R}r^2 \sin^2\theta \qquad (7.2.1)$$

$$\Gamma^1_{01} = \frac{\dot{R}}{R}$$

$$\Gamma^1_{11} = \frac{kr}{1-kr^2}$$

$$\Gamma^1_{22} = -r(1-kr^2)$$

$$\Gamma^1_{33} = -r(1-kr^2)\sin^2\theta$$

$$\Gamma^2_{02} = \frac{\dot{R}}{R}$$

$$\Gamma^2_{12} = \frac{1}{r}$$

$$\Gamma^2_{33} = -\sin\theta\cos\theta$$

$$\Gamma^3_{03} = \frac{\dot{R}}{R}$$

$$\Gamma^3_{13} = \frac{1}{r}$$

$$\Gamma^3_{23} = \cot\theta\,.$$

We now write the Ricci tensor, in the form:

$$R_{\mu\nu} = \Gamma^\sigma_{\mu\sigma,\nu} - \Gamma^\sigma_{\mu\nu,\sigma} + \Gamma^\rho_{\mu\sigma}\Gamma^\sigma_{\rho\nu} - \Gamma^\rho_{\mu\nu}\Gamma^\sigma_{\rho\sigma}\,. \qquad (7.2.2)$$

Applying the above formulae, we find the non-vanishing terms:

$$R_{00} = \frac{3\ddot{R}}{R}\;;$$

$$R_{11} = -(R\ddot{R} + 2\dot{R}^2 + 2k)/(1 - kr^2) \quad ;$$
$$R_{22} = -(R\ddot{R} + 2\dot{R}^2 + 2k)r^2 \quad ;$$
$$R_{33} = R_{22}\sin^2\theta \quad . \tag{7.2.3}$$

On the other hand, the perfect fluid energy-momentum tensor is given by:

$$T^\mu_\nu = (\rho + p)u^\mu u_\nu - p\delta^\mu_\nu. \tag{7.2.4}$$

We remember that the 4-velocity obeys the relation:

$$u^\mu u_\mu = 1 \,, \tag{7.2.5}$$

where $u^\mu \equiv \frac{dx^\mu}{ds}$, so that:

$$T \equiv T^\mu_\mu = (\rho + p) - 4p = \rho - 3p \quad . \tag{7.2.6}$$

In the comoving reference system:

$$u^\mu = \delta^\mu_0 \,, \tag{7.2.7}$$

$$u_\mu = g_{\mu\nu}\delta^\nu_0 = g_{\mu 0} = \delta^0_\mu \,, \tag{7.2.8}$$

$$T_{\mu\nu} = (\rho + p)\delta^0_\mu\delta^0_\nu - pg_{\mu\nu} \,, \tag{7.2.9}$$

$$T_{\mu\nu} - \tfrac{1}{2}Tg_{\mu\nu} = (\rho + p)\delta^0_\mu\delta^0_\nu - pg_{\mu\nu} - \tfrac{1}{2}(\rho - 3p)g_{\mu\nu} = (\rho + p)\delta^0_\mu\delta^0_\nu - \tfrac{1}{2}(\rho - p)g_{\mu\nu}. \tag{7.2.10}$$

We find now:

$$T_{00} - \tfrac{1}{2}Tg_{00} = \tfrac{1}{2}(\rho + 3p) \,, \tag{7.2.11}$$

$$T_{11} - \tfrac{1}{2}Tg_{11} = \frac{(\rho - p)R^2}{2(1 - kr^2)} \,, \tag{7.2.12}$$

$$T_{22} - \tfrac{1}{2}Tg_{22} = \tfrac{1}{2}(\rho - p)R^2r^2 \,, \tag{7.2.13}$$

$$T_{33} - \tfrac{1}{2}Tg_{33} = \tfrac{1}{2}(\rho - p)R^2r^2\sin^2\theta \,, \tag{7.2.14}$$

while the non-diagonal terms vanish.

The reader must become puzzled from the calculations made in (7.2.11) to (7.2.14). The explanation is given now. From Einstein's field equations:

$$G^{\mu\nu} \equiv R^{\mu\nu} - \tfrac{1}{2}Rg^{\mu\nu} = -\kappa T^{\mu\nu} \,, \tag{7.2.15}$$

we can derive the alternative expression,

$$R^{\mu\nu} = -\kappa\left(T^{\mu\nu} - \tfrac{1}{2}Tg^{\mu\nu}\right) \quad . \tag{7.2.16}$$

Hint:

Contract the mixed form of equation (7.2.15), $R^{\mu}_{\nu} - \frac{1}{2}R\delta^{\mu}_{\nu} = -\kappa T^{\mu}_{\nu}$; show that $R = -\kappa T$; and then verify the required result.

Let us return to Einstein's field equations; we are left with two equations:

$$\frac{3\ddot{R}}{R} = -\frac{1}{2}\kappa(\rho + 3p) \, , \qquad (7.2.17)$$

$$R\ddot{R} + 2\dot{R}^2 + 2k = \frac{1}{2}\kappa(\rho - p)R^2 \, . \qquad (7.2.18)$$

It is usual to write one equation for ρ alone, and other for p alone:

$$\kappa\rho = 3\frac{\dot{R}^2}{R^2} + \frac{3k}{R^2} \, , \qquad (7.2.19)$$

$$\kappa p = -\frac{2\ddot{R}}{R} + \frac{\dot{R}^2}{R^2} - \frac{k}{R^2} \, .$$

For the Lemaître models we add a density term $\frac{\Lambda}{\kappa}$, while we add a negative cosmic pressure $-\frac{\Lambda}{\kappa}$ thus finding:

$$\kappa\rho = 3H^2 + \frac{3k}{R^2} - \Lambda, \qquad (7.2.20)$$

$$\kappa p = -\frac{2\ddot{R}}{R} - H^2 - \frac{k}{R^2} + \Lambda \, . \qquad (7.2.21)$$

Note that we have defined Hubble's parameter, $H \equiv \frac{\dot{R}}{R}$, which determines the rate of expansion of the Universe. Notice also that the inclusion of the Λ constant as we did above, is justified by the results of section 3.5. Zeldovich calculated, from Quantum mechanical arguments, further that this procedure is equivalent to attributing an inherent density and negative pressure to the vacuum. This interpretation comes from displacing the Λ term from the left-hand side of the field equations to the r.h.s. In the l.h.s situation:

$$G_{\mu\nu} + \Lambda g_{\mu\nu} = -\kappa T_{\mu\nu} \, ,$$

we say that Λ is of geometric origin; however, if you write in the r.h.s.:

$$G_{\mu\nu} = -\kappa T_{\mu\nu} - \Lambda g_{\mu\nu} = -\kappa \overline{T}_{\mu\nu} \, ,$$

where we have defined an augmented energy tensor, which means that Λ is included in the new energy-tensor, as a property of matter, and not of the geometry. In both situations, we have the conservation of energy tensor:

$$T^{\mu}_{\nu;\mu} \equiv \overline{T}^{\mu}_{\nu;\mu} = 0 \, .$$

Likewise, we have for the l.h.s form:

$$[G^{\mu}_{\nu} + \Lambda g^{\mu}_{\nu}]_{;\mu} \equiv G^{\mu}_{\nu;\mu} = 0,$$

so that both forms are conserved.

From equations (7.2.20) and (7.2.21) we obtain:

$$\dot{\rho} + 3H(\rho + p) = 0 \,. \qquad (7.2.22)$$

The above equation could also have been derived by imposing the conservation of the energy momentum tensor:

$$T^{\mu\nu}_{;\nu} = 0. \qquad (7.2.23)$$

In fact, we get two equations:

$$(\rho u^{\mu})_{;\mu} + p u^{\mu}_{;\mu} = 0, \qquad (7.2.24)$$

and

$$(\rho + p) u^{\nu}_{;\mu} u^{\mu} = (g^{\mu\nu} - u^{\mu} u^{\nu}) p_{,\mu} \,. \qquad (7.2.25)$$

From (7.2.24),

$$\rho_{,\mu} u^{\mu} + (\rho + p)(u^{\mu}_{,\mu} + \Gamma^{\mu}_{\nu\mu} u^{\nu}) = 0 \,. \qquad (7.2.26)$$

If we now plug (7.2.7) into (7.2.26) we obtain (7.2.22), which is also called the continuity equation.

Now let us look to (7.2.25). For a comoving observer, we have:

$$u^{\mu} = \delta^{\mu}_{0}. \qquad (7.2.7)$$

With $\mu \neq 0$, we have:

$$(\rho + p) u^{\nu}_{;\mu} u^{\mu} = 0 \,,$$

because Krönecker's delta equals zero for $\mu \neq 0$, while:

$\frac{\partial p}{\partial r} \equiv \frac{\partial p}{\partial \theta} \equiv \frac{\partial p}{\partial \phi} \equiv 0$, because there is no pressure gradient, and the energy density and cosmic pressure only depend on time, due to homogeneity and isotropy. Hence,

$$\rho = \rho(t) \,;$$

$$p = p(t) \,; \quad \text{and}$$

$$(g^{\mu\nu} - \mu^{\mu} \mu^{\nu}) p_{,\mu} = 0$$

with $\mu \neq 0$.

For $\mu = 0$, as the metric tensor has zero covariant derivative, the l.h.s. of the (7.2.25) is equal to zero, leaving us with:

$$(g^{0\nu} - \mu^0 \mu^\nu) p_{,0} = 0 \, .$$

Because in general, $\frac{\partial p}{\partial t} \neq 0$, we have the above relation reduced to:

$$\delta^0_\nu - \delta^0_\nu = 0 \, .$$

On the other hand, we have already seen that the relation:

$$\mu^\nu_{;\mu} \mu^\mu = 0 \, ,$$

represents a geodesic, so that the fluid particles move along them; this is expected, indeed, since the particles are not subject to a pressure gradient to push them off geodesics ($\nabla p = 0$).

A useful identity may be derived by calculating the time derivative of the expression:

$$\dot{R}(t) \equiv H(t) R(t) \quad .$$

Then we find, by the product rule:

$$\ddot{R}(t) \equiv \dot{H}(t) R(t) + H(t) \dot{R}(t) \quad , \text{ or,}$$

$$\frac{\ddot{R}(t)}{R} \equiv \dot{H}(t) + H^2(t) \quad .$$

7.3. Constant Deceleration Parameter Models

As we have mention before, we define Hubble's parameter by:

$$H = \frac{\dot{R}}{R} \, .$$

Berman (1983) proposed that H varied with a power of $R = R(t)$, such as:

$$H = DR^{-m}, \qquad (7.3.1)$$

with D, m = constants.

For $m = 0$ we find H constant and by integration we obtain the law of variation of the scale-factor with time:

$$R = R_0 e^{Dt}, \qquad (7.3.2)$$

with R_0 = constant.

We find such law when dealing with exponential inflation, to be considered later.

If $m \neq 0$, we find, by integration, a possible law:

$$R = R(t) = (mDt)^{1/m} , \quad (7.3.3)$$

while from Einstein's field equations with $\Lambda \neq 0$, we find:

$$\rho = \frac{3D^2}{\kappa} R^{-2m} + \frac{3}{\kappa} kR^{-2} - \frac{\Lambda}{\kappa} , \quad (7.3.4)$$

and,

$$p = \frac{(2m-3)D^2}{\kappa} R^{-2m} - \frac{k}{\kappa} R^{-2} + \frac{\Lambda}{\kappa} . \quad (7.3.5)$$

Berman (1983), and Berman and Som (1989) found that these models have constant deceleration parameter $q = m - 1$ = constant , where we define:

$$q = -\frac{R\ddot{R}}{\dot{R}^2} . \quad (7.3.6)$$

The deceleration parameter was believed to be positive for the present Universe, that is why it was defined negatively; it is now thought that the present Universe is in fact in a process of acceleration, as we shall see later. An alternative definition, completely equivalent is:

$$q = -\frac{\ddot{R}}{RH^2} . \quad (7.3.7)$$

In order to show that the deceleration parameter is constant for the power-law (7.3.3) we take derivatives of $R(t)$ finding:

$$\dot{R} = DR^{-m+1} , \quad (7.3.8)$$

and,

$$\ddot{R} = -D^2(m-1)R^{-2m+1} . \quad (7.3.9)$$

We now plug (7.3.8) and (7.3.9) into (7.3.6) finding:

$$q = m - 1 = const . \quad (7.3.10)$$

We now obtain for Hubble's parameter the following law:

$$H = (mt)^{-1} . \quad (7.3.11)$$

The present author has been apparently the first one to publish the above formula, or its equivalent form:

$$H = [(1+q)t]^{-1} . \quad (7.3.12)$$

In other words, the age of the Universe would be given by $\frac{H^{-1}}{m} = \frac{H^{-1}}{q+1}$.

Astronomers have found that:

$$H_0^{-1} \cong 1.8\text{x}10^{10} years , \quad (7.3.13)$$

where the subscript zero means "present value".

If we now calculate $\dddot{R}$, we find:

$$\dddot{R} = -D^2(m-1)(1-2m)\dot{R}R^{-2m} , \qquad (7.3.14)$$

so that the variation of deceleration parameter is given by the pure number (adimensional number), suggested by Peter Landsberg(1983), who also made references to q – constant models in terms of rough calculations:

$$Q = \frac{R^2\dddot{R}}{\dot{R}^3} = (2m-1)(m-1) . \qquad (7.3.15)$$

From (7.3.15) with (7.3.9) we find:

$$Q = 2q^2 + q . \qquad (7.3.16)$$

We conclude that the $Q-$parameter is also constant, for constant deceleration parameter models.

By plugging (7.3.8) and (7.3.9) into (7.3.4) and (7.3.5), we can find the time dependence for the energy density and cosmic pressure:

$$\rho = \left(\frac{3}{\kappa m^2}\right)t^{-2} + \frac{3k}{\kappa(m^2D^2)^{1/m}}t^{-2/m} - \frac{\Lambda}{\kappa} , \qquad (7.3.17)$$

and,

$$p = \left[\frac{(2m-3)}{\kappa m^2}\right]t^{-2} - \frac{k}{\kappa(m^2D^2)^{1/m}}t^{-2/m} + \frac{\Lambda}{\kappa} . \qquad (7.3.18)$$

It can be checked that $k = 0$ and $m = \frac{3}{2}$ and $\Lambda = 0$ gives the so called "Friedmann flat model" where $\rho = \rho_{critical} = 3H^2/\kappa$. This critical density is one of the most important parameters in Cosmology (See Section 7.7).

7.4. Cosmological Red-Shift

We shall now show that, for a light wave emitted in cosmic time t_E , when the scale-factor

is $R(t_E)$, and which is received at present time t_0 , when the scale-factor is $R(t_0) > R(t_E)$ – (in other words, the Universe is expanding) – the wavelength λ_0 received is larger than the one emitted, λ_E :

$$\frac{\lambda_0}{\lambda_E} = \frac{R(t_0)}{R(t_E)} > 1 . \qquad (7.4.1)$$

The spectrum of such light shows a RED-SHIFT.

Consider a wave front of an electromagnetic emission at the initial time t_E . For a radial geodesic,

$$d\theta = d\phi = 0 .$$

The radial coordinate being "r" for the emitting "galaxy", we have from Robertson-Walker's metric, with the $ds = 0$,

$$-\frac{R^2(t)}{(1+\frac{k}{4}r^2)^2}dr^2+dt^2=0 \ . \qquad (7.4.2)$$

We expect for a far-away galaxy, that the periods dt_E , and dt_0 , of the emitted and received photon, are much smaller than the total time of the trajectory, $[t_0-t_E]$. From equation (7.4.2), we find now that:

$$\int_{t_E}^{t_0}\frac{dt}{R(t)}=-\int_{r}^{0}\frac{dr}{1+\frac{kr^2}{4}} \ , \qquad (7.4.3)$$

and also,

$$\int_{t_E+dt_E}^{t_0+dt_0}\frac{dt}{R(t)}=-\int_{r}^{0}\frac{dr}{1+\frac{kr^2}{4}} \ . \qquad (7.4.4)$$

We now recall the linearity of the definite integral operation:

$$\int_{t_E+dt_E}^{t_0+dt_0}\equiv\int_{t_E+dt_0}^{t_E}+\int_{t_E}^{t_0}+\int_{t_0}^{t_0+dt_0} \ , \qquad (7.4.5)$$

so that,

$$\int_{t_E+dt_E}^{t_0+dt_0}\frac{dt}{R(t)}=-\frac{dt_E}{R(t_E)}+\int_{t_E}^{t_0}\frac{dt}{R(t)}+\frac{dt_0}{R(t_0)}=\int_{t_E}^{t_0}\frac{dt}{R(t)} \ , \qquad (7.4.6)$$

and, taking this result into (7.4.3) and (7.4.4) we find:

$$-\frac{dt_E}{R(t_E)}+\frac{dt_0}{R(t_0)}=0 \ , \qquad (7.4.8)$$

which takes us back to (7.4.1), where we have already included the result that we know that $R(t_0)>R(t_E)$, because:

$$1+\left(\frac{\Delta\lambda}{\lambda_E}\right)\equiv 1+z=\frac{R_0}{R_E}=\frac{R(t_0)}{R(t_E)}>1 \ . \qquad (7.4.9)$$

Notice that $z>0$, for a red-shift ($\Delta\lambda>0$) and that we have used the standard equation:

$$\lambda=cT \ ,$$

where T stands for period, and λ is the wavelength.

It was the American astronomer Edwin Hubble, the first, or one of the first, to discover the expansion of the Universe by analyzing the spectra of light from nearby galaxies. He found Hubble's law, under which, for nearby galaxies,

$$z\cong H_0 d_L \ , \qquad (7.4.10)$$

where $H_0=H(t_0)$ = Hubble's "constant", and d_L stands for the "luminosity" distance of the galaxy as seen from the earth. In next section, we shall show what we mean by d_L which is roughly the distance, from the earth, of the galaxy; while we shall also show how to obtain Hubble's law from Roberston-Walker's metric.

7.5. Hubble's Law

If the physical distance is "d" between two galaxies, which are comoving, and are separated by coordinate distance d_0 , we find:

$$d = R(t)d_0 = d(t) . \qquad (7.5.1)$$

In fact, the above is the translation of what we mean by the expanding metric.

The relative speed between both galaxies, is, then:

$$v = \frac{d}{dt}[d(t)] = \dot{R}d_0 = \dot{R}[\frac{d(t)}{R(t)}] = H(t)d(t) . \qquad (7.5.2)$$

This speed is interpreted, by forgetting peculiar random speeds of galaxies, as a recessional speed. So, Hubble measured red-shifts and attributed them to recessional speeds, that are as much larger as its luminosity distance is from us.

For a comoving source and observer, with distance d apart, with $d << 1$, we know that:

$$1 + z \cong \frac{R_0}{R_E} . \qquad (7.5.3)$$

We now expand R_0 about R_E keeping up to the first derivative in Taylor's series:

$$R_E \cong R_0 - \dot{R}_0[\Delta t_1] , \qquad (7.5.4)$$

where,

$$\Delta t_1 \equiv t_0 - t_1 \equiv \frac{d}{c} , \qquad (7.5.5)$$

and where d is loosely called the distance between the two galaxies. Then,

$$z = \frac{R_0}{R_E} - 1 \cong \frac{1}{1 - H_0 d} - 1 \cong H_0 d \qquad \text{q.e.d.}$$

7.6. Cosmological Model for Large Lambda

Let us recall the two equations for Cosmology, one for energy density, the other for cosmic pressure (7.2.20) and (7.2.21):

$$\kappa\rho = 3H^2 + \frac{3k}{R^2} - \Lambda , \qquad (7.6.1)$$

$$\kappa p = -\frac{2\ddot{R}}{R} - H^2 - \frac{k}{R^2} + \Lambda . \qquad (7.6.2)$$

Suppose that Λ is the driving "force" in the Universe, so that:

$$\kappa\rho << \Lambda , \qquad (7.6.3)$$

$$\Lambda >> R^{-2} > 0 , \qquad (7.6.4)$$

$$\Lambda >> \kappa p . \qquad (7.6.5)$$

Note that in the ρ and p , terms, of this Section, the Λ – contribution is not present; what they represent, is just the real matter fields.

From (7.6.1) we find:

$$3H^2 \cong \Lambda \,, \qquad (7.6.6)$$

or,

$$\frac{\dot{R}}{R} \cong \sqrt{\frac{\Lambda}{3}} \,. \qquad (7.6.7)$$

On integrating,

$$R \cong R_0 e^{\sqrt{\frac{\Lambda}{3}}\, t} \,, \qquad (7.6.8)$$

where $R_0 = R(t=0) = \text{constant}$.

Equation (7.6.2) is then fulfilled identically. This scale-factor is representative of exponential inflation (see next Section). From next Chapter, we shall see that for Planck's Universe, in the very early Universe, which stands at the limit between the validity of Classical Physics and Quantum Physics, Λ is in fact the largest driving cause for the expansion of the Universe. From Einstein's equations as above, we can predict thus, that inflationary expansion must occur in the very early Universe. It could be argued that, because for the present Universe, the Λ constant is responsible for a very large amount of the energy density of the present Universe ($\frac{\Lambda}{\kappa}$ is very large), the present Universe could be also in an exponential inflationary phase. For this phase, the deceleration parameter is $q = -1$, as can be checked by plugging the inflationary law into the definition for q (see Section 7.3). In fact, we can imagine that $-1 \leq q < 0$. So that, the Universe is accelerating, as recent Supernovae observations have revealed. However, it seems that $\frac{\Lambda}{\kappa}$ is not the only driving "force" in the present Universe (that is why q is not equal to -1 ; by the same token, we could have, or not, the equation of state , $p \neq -\frac{1}{3}\rho$. (See Section 9.7).

It seems however, that the Λ value for the very early Universe should be 10^{120} times larger than its value for the present Universe. The Λ value today is limited by the fact that Schwarzschild's solution for the solar system is more or less accurate with $\Lambda \approx 0$, according to astronomers.

We mention here that the Schwarzschild's metric with a Λ constant included is given by (Tolman, 1934):

$$ds^2 = \frac{-dr^2}{1-\frac{2m}{r}-\frac{\Lambda}{3}r^2} - r^2(d\theta^2 + \sin^2\theta d\phi^2) + c^2(1 - \frac{2m}{r} - \frac{\Lambda}{3}r^2)dt^2 \,. \qquad (7.6.9)$$

7.7. Inflation

A closely related idea, to the large lambda model considered above, is "inflation". As we shall see in next Section, and further in next Chapter, one is led to believe that, at least, for the very early Universe, an exponential scale-factor driven by a large lambda, has in fact occurred. From Grand Unified Theories (GUT's) which tell that the unified field covered by the nuclear strong and weak, and the electromagnetic, interactions, had "afterwards", under the action of a scalar field, broken its "symmetry". Then, the energy of the vacuum (V_0) was driving the Universe, say with $\rho \cong V_0$. This is tantamount to choosing a model with large Λ , where according to last Section,

$$H = \sqrt{\tfrac{\kappa}{3}V_0} = \sqrt{\tfrac{\Lambda}{3}} .$$

GUT's time, according to Particle Physics is about 10^{-38} seconds after the "birth" of the Universe ($t = 0$), which is very close to Planck's time, 10^{-43} seconds, which is the minimum time for the failure of Classical Physics and the entrance of Quantum Physics in the scenario (see next Section), when we go back in time.

Among the problems solved by inflation we list the "flatness" and the "horizon" problems plus "monopoles" production in the Universe. The "flatness" problem is basically how one can explain that the Universe today is so close to being flat. For instance, the visible matter energy density, would point to a magnitude of the energy density only one hundred times less the critical density for the Universe, which is a "close" number. The critical density is the density for which $k = 0$ in the Robertson-Walker's metric when the Universe is pressureless (Friedmann models). The density parameter is defined for the total energy density as $\Omega = \frac{\kappa\rho}{3H^2}$, so that for the critical density, $\Omega = 1$. Analogously, we define for the cosmological constant, a density parameter, $\Omega_\Lambda = \frac{\Lambda}{3H^2}$, while for the energy density of matter we define $\Omega_m = \frac{\kappa\rho_m}{3H^2}$, where ρ_m denotes the energy density of the matter in the Universe. It should be noted that experimentally $\Omega_\Lambda \sim 0.7$ and then if $k = 0$, we shall have $\Omega_m \sim 0.3$. From (7.6.1) , it follows that the energy density is the "critical" one , when $k = \Lambda = 0$.

Returning to the flatness problem, this is caused because it would have required a very fine tuning of the initial conditions for the Universe in order to obtain, today, a flat Universe.

The horizon problem is defined by the question of how can we explain why different disconnected parts of the Universe have similar or same physical properties. This could not have been achieved by power law scale factors, that would expand faster than the speed of light traveling from the one part to the other in the Universe, sufficiently far away. The rapidly exponential growth solves this problem, because it allows the early radius of space to be much smaller than the horizon distance $r_h = ct$.

The flatness problem is also solved by the rapid expansion caused by an inflationary phase because it turned the Universe more flat quickly. The best picture for this phenomenon is to imagine the Universe as the surface of a tridimensional balloon: as the balloon inflates the surface becomes more flat.

The monopoles production in the very early Universe is a problem, for no trace of such monopoles, required by Particle Physics theories to have been produced in the very early Universe, are found in the present Universe. The explanation given by inflationary theorists is that there was a finite amount of monopoles before inflation; with the rapidly expansion caused by inflation, they were diluted, so that no trace is found today.

7.8. Planck's Universe and the Λ-problem

The limit for Classical Cosmology becomes evident when one tries to combine the constants of Physics into dimensional quantities. For Classical Physics the constants are two, namely, the velocity of light in vacuum, c, and Newton's gravitational constant, G: one represents electromagnetism, and the other, the gravitation. For Quantum Physics there is another constant, (Planck's), $\bar{h} = h/2\pi$.

When we combine all of them, we obtain a length dimension, the Planck's length:

$$L_{Pl} = \left(\frac{\hbar G}{c^3}\right)^{\frac{1}{2}} \cong 1.6\text{x}10^{-33} cm \; .$$

Analogously, we find Planck's time (t_{Pl}) and Planck's mass (M_{Pl}):

$$t_{Pl} = \left(\frac{\hbar G}{c^5}\right)^{\frac{1}{2}} \cong 5\text{x}10^{-44} \sec \; ,$$

$$M_{Pl} = \left(\frac{\hbar c}{G}\right)^{\frac{1}{2}} \cong 2\text{x}10^{-5} g \; .$$

We could try to find a value for the Λ - constant, by constructing a combination of the constants yielding a length to the minus second power. This is:

$$\Lambda_{Pl} \cong L_{Pl}^{-2} \cong 10^{66} cm^{-2} \; ,$$

regarded as the Cosmological constant for Planck's Universe.

On the other hand, though we can keep the above values, for Planck's Universe, which represents the limit between Quantum gravity theory and Classical General Relativity, we know that the above value for Λ_{Pl} can not exist for the present Universe. In fact, it is estimated from supernovae high red-shift observations that:

$$\Lambda_0 \cong 10^{-56} cm^{-2} \; .$$

The above experimental result, matches with the upper limit for the "constant", which could possibly enter in the Schwarzschild's metric with Λ (see equation 7.6.9), without disturbing the data gathered from solar system's observations, when compared with the same equation with $\Lambda = 0$.

The ratio:

$$\Lambda_{Pl}/\Lambda_0 \approx \frac{10^{66}}{10^{-56}} = 10^{122} \; ,$$

constitutes the most serious problem of contemporary Cosmology, being called the Cosmological Constant Problem or briefly the lambda-problem.

We have seen that we have two distinct evaluations for the Λ constant: the first is the very early Universe Λ estimated as the second minus power of the Planck's length; the second is the estimated value by means of high red-shift supernovae observations, which is 10^{-120} times smaller, for the present Universe. This is really a problem because the constant is not a constant!!!. A possible solution for this enigma, as advanced by several authors including the present author (Berman and Som, 1990 a ; Berman, 1990 ; Berman, 1991 a ; 1991 b ; 1992 b ; 1992 ; 1994 ; Berman and Trevisan, 2002 a ; 2001 a ; 2001 b ; 2001 c ; Berman, Som and Gomide, 1989), following a hint by the Brazilian physicist Orfeu Bertolami (1986), was to consider that we have not a constant, but really a Λ-term varying with a minus square power of the age of the Universe:

$\Lambda \propto t^{-2}$ (Bertolami-Berman solution).

In fact,

$$\left(\frac{t_{GUT}}{t_0}\right)^{-2} \cong \left(\frac{10^{-34}\,\mathrm{sec}}{10^{10} years}\right)^{-2} \cong 10^{120} .$$

This resembles approximately the necessary ratio. On the other hand, another suggested Λ - term variation was given by Chen and Wu (1990), by dimensional analysis, involving constants c, G, and h, when Planck´s constant would not play any role.

$$\Lambda \propto R^{-2} .$$

In fact, consider the ratio:

$$\left(\frac{L_{Pl}}{R_0}\right)^{-2} \cong \left(\frac{10^{-33} cm}{10^{28} cm}\right)^{-2} \cong 10^{122} .$$

An exhaustive paper, containing models with a Λ - term with both kinds of variation and even with powers different than the second power was studied by Overduin and Cooperstock (1998).

It must be remarked that the first paper was Bertolami's and considered the Brans-Dicke formalism which is an alternative theory of gravity (see next Chapter). Other similar treatments are due to Abdel-Rahman (1990) , Özer and Taha (1986), and others. Special mention is to a paper by Arbab I. Arbab (1997).

The cosmological constant has been considered in the context of Quantum Field Theories, Kaluza-Klein penta-dimensional theories, quantum gravity, Grand-Unified Theories and supergravity. It is believed, from gauge theories of elementary particles, that the cosmological constant is related to the mass of the Higgs boson. As a dynamical quantity the cosmological term has been studied as anything that contributed to the energy density of the vacuum. Potential sources were identified for fluctuating vacuum energies including scalar fields (Dolgov, 1983 ; Abbot, 1985 , etc); tensor fields (Hawking, 1984 ; Banks, 1984 ; Brown and Teitelboim, 1987 ; Dolgov, 1997); non-local effects (Linde, 1988 ; Banks, 1988); wormholes (Coleman, 1988); inflation phenomena (Tsamis and Woodart, 1995 ;

Brandenberger and Zhinitsky, 1997); and also with cosmological perturbations (Abramo, Brandenberger and Mukhanov, 1997); etc.

The most popular solutions for a decaying Λ are picked in Table I of Overduin and Cooperstock (1998) definitive paper; Λ may be thought to vary not only with t^{-2} but with others powers of t ; in some other models it varies either with H (Hubble's parameter), q (deceleration parameter), or the scale factor, temperature, or in combinations of them.

We notice that the natural spin associated with Planck´s Universe is given by Planck´s constant, and hints that the Universe has an angular momentum.

7.9. Raychaudhuri's Equation and an Alternative Form

A.K. Raychaudhuri (1979) derived a most important equation which takes his name, which is applied in the study of the so-called "singularity theorems"; though we shall not delve in detail into those theorems, we shall use this equation in connection with the study of density perturbations, to be undertaken in next section.

From the definition of the Riemann-Christoffel tensor, we have, for any vector v^{γ},

$$v^{\mu}_{;\alpha;\beta} - v^{\mu}_{;\beta;\alpha} = R^{\mu}_{\gamma\beta\alpha} v^{\gamma} . \tag{7.9.1}$$

From (7.9.1), and the definition of the Ricci tensor, we find:

$$v^{\mu}_{;\alpha;\mu} v^{\alpha} - v^{\mu}_{;\mu;\alpha} v^{\alpha} = R_{\gamma\alpha} v^{\alpha} v^{\gamma} . \tag{7.9.2}$$

If v^{μ} is a unit vector, we define the following terms:

i) Expansion scalar: $\theta \equiv v^{\mu}_{;\mu}$

ii) Acceleration vector: $\dot{v}^{\mu} \equiv v^{\mu}_{;\alpha} v^{\alpha}$

iii) Vorticity tensor: $\omega_{\alpha\beta} \equiv \frac{1}{2}\left[v_{\alpha;\beta} + \dot{v}_{\beta} v_{\alpha} - (v_{\beta;\alpha} + \dot{v}_{\alpha} v_{\beta})\right]$

iv) Shear tensor: $\sigma_{\mu\nu} \equiv \frac{1}{2}\left[v_{\mu;\nu} + v_{\nu;\mu} - (\dot{v}_{\mu} v_{\nu} + \dot{v}_{\nu} v_{\mu})\right] - \frac{1}{3} v^{\alpha}_{;\alpha}\left(g_{\mu\nu} - v_{\mu} v_{\nu}\right)$

From the above definitions, and the unitarity of v^{ν} , we have:

a) $\dot{v}^{\mu} v_{\mu} = 0$

b) $\omega_{\alpha\beta} v^{\alpha} = \omega_{\alpha\beta} v^{\beta} = 0$

(The vorticity tensor is antisymmetric and orthogonal to v^{α}).

c) $\sigma_{\mu\nu} v^{\nu} = \sigma_{\mu\nu} v^{\mu} = 0$

(The shear tensor is symmetric, tracefree and orthogonal to v^{α}).

Now, with the above definitions, we get:

$$v_{\mu;\nu} = \sigma_{\mu\nu} + \omega_{\mu\nu} + \frac{1}{3}\theta\left(g_{\mu\nu} - v_\mu v_\nu\right) + \dot{v}_\mu v_\nu, \tag{7.9.3}$$

and, from (7.9.2):

$$\left(v^\mu_{;\alpha} v^\alpha\right)_{;\mu} - v^\mu_{;\alpha} v^\alpha_{;\mu} - \theta_{,\alpha} v^\alpha = R_{\nu\alpha} v^\alpha v^\nu \,. \tag{7.9.4}$$

From (7.9.3) and (7.9.4), we can write:

$$\dot{v}^\mu_{;\mu} - 2(\sigma^2 - \omega^2) - \frac{1}{3}\theta^2 - \theta_{,\alpha} v^\alpha = R_{\nu\alpha} v^\alpha v^\nu \,, \tag{7.9.5}$$

where:

$$\sigma^2 \equiv \frac{1}{2}\sigma_{\mu\nu}\sigma^{\mu\nu} \,, \tag{7.9.6}$$

$$\omega^2 \equiv \frac{1}{2}\omega_{\mu\nu}\omega^{\mu\nu} \,. \tag{7.9.7}$$

When v^ν is timelike, $\sigma^2 \geqslant 0$ and $\omega^2 \geqslant 0$. The equalities only stem from the vanishing of all the components of $\sigma_{\mu\nu}$ and $\omega_{\mu\nu}$ respectively.

By plugging into (7.9.5), the Einstein's field equations and the expression for a perfect fluid energy-momentum tensor, we find:

$$\theta_{,\alpha} v^\alpha + \frac{1}{3}\theta^2 - \dot{v}^\alpha_{;\alpha} + 2(\sigma^2 - \omega^2) + \frac{\kappa}{2}(\rho + 3p) - \Lambda = 0 \,, \tag{7.9.8}$$

and, also,

$$\dot{v}^\gamma = p_{,\sigma} \frac{(g^{\sigma\gamma} - v^\sigma v^\gamma)}{(\rho + p)} \,. \tag{7.9.9}$$

With the definition,

$$H = \frac{\dot{R}}{R} = \frac{\theta}{3} \tag{7.9.10}$$

we obtain:

$$q \equiv -\frac{\ddot{R}R}{\dot{R}^2} = H^{-2}\left[\frac{\kappa}{6}\sigma_w + \frac{2}{3}\sigma^2 - \frac{2}{3}\omega^2 - \frac{1}{3}\dot{v}^\mu_{;\mu}\right] \tag{7.9.11}$$

In fact, the Raychaudhuri's equation (7.2.18) becomes,

$$\dot{\theta} + \frac{\theta^2}{3} - \dot{v}^\mu_{;\mu} + 2(\sigma^2 - \omega^2) + \frac{1}{2}\kappa(\rho + 3p) - \Lambda = 0 \tag{7.9.12}$$

where σ stands for the shear tensor magnitude, $\dot{v}^\mu_{;\mu}$ the acceleration, ω the spin vector magnitude, ρ the energy density, p the cosmic pressure, Λ the cosmological constant, and θ the dilatation. The Robertson-Walker's scale factor R and Hubble's parameter H are related by :

$$H = \frac{\dot{R}}{R} = \frac{\theta}{3} \tag{7.9.10}$$

It can be shown that R stands for the average "radius" of the Universe. For instance, if we had Bianchi I metric,

$$ds^2 = dt^2 - A^2dx^2 - B^2dy^2 - C^2dz^2 \tag{7.9.13}$$

and

$$R^6 = A^2B^2C^2 \tag{7.9.14}$$

Now, we go back to the Raychaudhuri's equation (7.9.12) and find (Narlikar, 1983):

$$\frac{3\ddot{R}}{R} = 2\omega^2 - 2\sigma^2 + \dot{v}^{\mu}_{;\mu} - \frac{1}{2}\kappa(\rho + 3p) + \Lambda \tag{7.9.15}$$

If we define $\Omega = \frac{\rho}{\rho_{crit}}$ (7.9.16)

and

$$q = -\frac{\ddot{R}R}{\dot{R}^2} \tag{7.9.17}$$

we find:

$$3H^2\left(\frac{\Omega}{2} - q\right) = 2\omega^2 - 2\sigma^2 + \dot{v}^{\mu}_{;\mu} - \frac{3}{2}\kappa p + \Lambda \tag{7.9.18}$$

The last equation is the desired alternative form for Raychaudhuri's equation in terms of the parameters of the theory. Berman (1994 a) seems to have been the first one to publish relation (7.9.18). This equation tells that the effects of shear, vorticity, acceleration, pressure, and cosmological constant may be magnified with the age of the Universe. In fact, we have already shown before, that an approximate relation, which is exact for constant deceleration parameters, is the following:

$$H \cong \frac{1}{(1+q)\,t}\,, \tag{7.9.19}$$

so that,

$$3\left(\frac{\Omega}{2} - q\right) \cong \left[2\omega^2 - 2\sigma^2 + \dot{v}^{\mu}_{;\mu} - \frac{3}{2}\kappa p + \Lambda\right] t^2\ (1+q) \tag{7.9.20}$$

If the left hand side of the above equation is constant, then we might think of the following possible solution:

$$\omega = At^{-1}$$

$$\sigma = Bt^{-1}$$
$$\dot{v}^{\mu}_{;\mu} = Ct^{-2}$$
$$\Lambda = Dt^{-2}$$
$$p = Et^{-2}$$

where $A, B, C, D,$ and E are constants. Then, the following relation is valid:

$$3\left(\frac{\Omega}{2} - q\right) \cong \left[2A^2 - 2B^2 + C - \frac{3}{2}\kappa E + D\right]\ (1+q) \tag{7.9.21}$$

The above equation (7.9.18), deserves attention from theoreticians.

7.10. Geodesics for Photons, Particles, and Tachyons in Robertson-Walker's Cosmology

A very clear study of geodesics in the isotropic Universe was undertaken by Chaliasos (1987), and by Berman (1997). From the analytic form of the geodesics in the Robertson-Walker's metric, and also found the solutions for the closed, flat and open Friedmann models, in the dust and radiation phases, with a null cosmological constant. Here, we shall study geodesics in the inflationary phase, and also with a solution for the general case of a perfect gas equation of state:

$$p = \gamma\rho \qquad (\ \gamma = \text{const}\ ) \tag{7.10.1}$$

We restrict to flat Universes, with constant deceleration parameter models. First we detail Chaliasos approach.

Introduction to the Problem

Consider Robertson-Walker's metric in the alternative form (Weinberg, 1972):

$$ds^2 = c^2dt^2 - R^2(t)d\Omega^2 = dt^2 - dl^2\ .$$

Here, $R(t)$ stands for the scale-factor, while $l = l(r,\theta,\phi)$ is the space component of the metric. We assume that the origin of space-coordinates are on the orbit of the moving particle where, for symmetry reasons we have $\theta = \phi =$ constant .

The speed, energy, and linear momentum, of a particle, for such Universe is defined by :

$$v = \frac{dl}{dt} = R(t)\frac{dr}{dt}\ ; \qquad (d\theta = d\phi = 0),$$

$$E = \frac{mc^2}{\sqrt{1-\left(\frac{v}{c}\right)^2}};$$

and,

$$p = \frac{m\,v}{\sqrt{1-\left(\frac{v}{c}\right)^2}}$$

$$\vec{p} \equiv \left(p_r, p_\theta, p_\phi\right);$$

$$p^2 = -g^{ij}p_ip_j = \frac{1}{R^2}p_r^2; \quad (p_\theta = p_\phi = 0)\ .$$

$$p = \frac{p_r}{R}\ .$$

In order to accommodate the above treatment for massive particles, zero-rest-mass photons, and tachyons, which have speeds greater than the speed of light, we substitute m^2 by

λm^2 , where $\lambda = -1$ for tachyons, $\lambda = +1$ for normal particles, and $\lambda = 0$ for photons. From the above relations, taking care of the constant λ , we find:

$$\frac{v}{c} = \left\{ \sqrt{1 + \frac{\lambda m^2 c^2 R^2}{p_r^2}} \right\}^{-1}.$$

We also obtain,

$$\frac{v}{c} = \frac{R}{c}\frac{dr}{dt}.$$

From the last two equalities, we can integrate the radial coordinate in terms of cosmic time t , finding:

$$r = \int dr = \pm \int \left\{ \sqrt{1 + \frac{\lambda m^2 c^2 R^2}{p_r^2}} \right\}^{-1} R^{-1} c \, dt \, .$$

This is the result of Chaliasos

Null Geodesics

The general equation for null geodesics is:

$$r = \pm \int \frac{dt}{R} \tag{7.10.2}$$

For exponential inflation,

$$R = R_0 e^{Ht} \tag{7.10.3}$$

with

$$p = -\rho \tag{7.10.4}$$

or $\gamma = -1$, we find:

$$r = \pm \frac{e^{-Ht}}{R_0 H} + r_0 \qquad (r_0 = \text{const}) \tag{7.10.5}$$

We interpret this result as meaning that a photon in the inflationary phase will be approximately comoving, and

$$r \cong r_0 \tag{7.10.6}$$

In other words, photons are "localized", in the inflationary scenario.

For power-law solutions and constant deceleration parameter:

$$\begin{gathered} R(t) = (mDt)^{1/m} \\ \gamma = \tfrac{1}{3}(2m-3) \\ k = 0 \\ m = q+1 \end{gathered} \tag{7.10.7}$$

From (7.10.2), we find, for null geodesics, if $q \neq 0$,

$$r = \pm\left[\left(1-\tfrac{1}{m}\right)(mD)^{1/m}\right]^{-1} t^{1-1/m} + r_0 \qquad (7.10.8)$$

This is an increasing function of t for $m > 1$. When $t = 0$, $r = r_0$. In the limit $t \to \infty$, we have $r \to \infty$, as expected for an accelerated Universe ($m < 1$). We have a photon at infinity when $t = 0$ and then when $t \to \infty$ we find that $r \to r_0$ (comoving photon).

If $m = 1$, we find:

$$r = \pm(D)^{-1/2} \ln t + r_0 \qquad (7.10.9)$$

In order to have a meaningful result, in this case we must choose the plus sign, and then, when $t \to \infty$, also $r \to \infty$. We have to adjust r_0, so that for $t \to t_{Planck} \cong 10^{-43}$ sec , $r \to 0$ (comoving). Of course, since we are in a Classical environment, for $t < 10^{-43}$ sec , our model does not apply.

Geodesics for Particles and Tachyons in a flat Universe

Suppose that particles and tachyons live in a $p = \alpha\rho$ Universe with $\alpha =$ const . This happens for power-law constant deceleration parameter flat Universes with:

$$\alpha = \tfrac{2m-3}{3} \qquad (7.10.10)$$

or for the exponential inflationary $m = 0$ case. By varying m, we vary α in (7.10.10) and then we sweep the general perfect gas equation of state cases.

Consider first the inflationary case. The geodesics is given by:

$$r = \pm\int \{R_0 e^{Dt}[1 + \lambda\alpha'^2 R_0 e^{2Dt}]^{1/2}\}^{-1} dt \qquad (7.10.11)$$

where $\alpha' = m'^2c^2/\mathbf{p}_r^2 =$ const. Here, m' is the rest-mass, and $\mathbf{p}_r$ stands for the radial momentum, and $\lambda = +1$ for material particles and $\lambda = -1$ for tachyons. With the numerical data available for the inflationary phase, and R_0 of order $10^{-33} cm$ (Planck's length), we verify that the above integral reduces to the corresponding null geodesics case:

$$r \cong \pm\int R_0^{-1} e^{-Dt} dt \cong \text{const.} \qquad (7.10.12)$$

(comoving or "localized" particle).

Let us now consider the $m \neq 0$ cases. We have:

$$r = \pm\int \{(mDt)^{1/m}[1 + \lambda\alpha'^2 (mDt)^{2/m}]^{1/2}\}^{-1} dt \qquad (7.10.13)$$

For $\lambda = +1$ (particles) and $m = 1$, we find:

$$r = \mp D^{-1} \ln\left[\frac{\sqrt{t^2+1/K} + K^{-1/2}}{t}\right] \qquad (7.10.14)$$

where $K = \alpha'^2 D^2$.

When $\lambda = -1$ (tachyons), and $m = 1$, we have:

$$r = \mp D^{-1} \ln \left[\frac{\sqrt{K^{-1} - t^2} + K^{-1/2}}{t} \right] \tag{7.10.15}$$

For $m = 2$ we have Chaliassos parametric analytic form:

$$r = \pm \alpha'^{-1} \{ \alpha' y + \sqrt{1 + \alpha'^2 y} \} \pm \text{const} \tag{7.10.16}$$

which is valid for particles ($\lambda = +1$) and where

$$dy = R^{-1} dt \tag{7.10.17}$$

For tachyons:

$$r = \pm \alpha'^{-1} \arcsin(\alpha' y) \tag{7.10.18}$$

It is expected that other cases yield similar results, through numerical computations.

7.11. Energy of Robertson-Walker's Universe

The pioneer works of Berman (1981) , Nathan Rosen(1994), Cooperstock and Israelit(1995), showing that the energy of the Universe is zero, by means of calculations

involving pseudotensors, and Killing vectors, respectively, are here given a more simple approach. We shall show that the energy of the Robertson-Walker's Universe is zero, that this Universe is Machian, and it lies inside a black-hole, thus hinting altogether to a solution of the cosmological constant problem (Berman,1981; 2006;2006b;2007;2009).

Consider Minkowski's metric,

$$ds^2 = dt^2 - \left[dx^2 + dy^2 + dz^2 \right]. \tag{7.11.1}$$

This is an empty Universe, except for test particles. We agree that its total energy is zero (Weinberg, 1972).

Now consider the expanding flat metric:

$$ds^2 = dt^2 - R^2(t) \left[dx^2 + dy^2 + dz^2 \right] . \tag{7.11.2}$$

Here, $R(t)$ is the scale-factor. At any particular instant of time, $t = t_0$, we may define new variables, by the reparametrization,

$$dx'^2 \equiv R^2(t_0) dx^2 , \tag{7.11.3}$$

$$dy'^2 \equiv R^2(t_0) dy^2 , \tag{7.11.4}$$

$$dz'^2 \equiv R^2(t_0)\, dz^2 \,, \tag{7.11.5}$$

$$dt'^2 \equiv dt^2 \,.$$

Then,

$$ds^2 = dt'^2 - \left[dx'^2 + dy'^2 + dz'^2\right]. \tag{7.11.6}$$

The energy of this Universe is the same as Minkowski's one, namely, $E = 0$. We remember that in order to obtain the energy of a physical system, we have to consider a definite fixed instant of time. As t_0 has been chosen arbitrarily , our result is time-independent.

Consider now the metric:

$$ds^2 = dt^2 - \frac{R^2(t)}{\left[1+\frac{kr^2}{4}\right]^2}\left[dx^2 + dy^2 + dz^2\right] . \tag{7.11.7}$$

Here, $k = 0$ yields the flat case, already studied. When $k = \pm 1$, we have finite closed or infinite open Universes.

We want to calculate its energy. We are allowed to choose the way into making the calculation, so we consider $k = \pm 1$ (finite / open Universes).We reparametrize the metric:

$$dx'^{i2} \equiv \frac{R^2(t_0)dx^{i2}}{\left[1+k\frac{r^2}{4}\right]^2} \qquad (\, i = 1,2,3\,) \tag{7.11.8}$$

The energy is, again,of zero value. As above, this is also a time-independent zero- total energy.

7.12. The Zero-Total Energy Density of the Universe

Consider the energy-density Friedman-Robertson-Walker equation,

$$\kappa\rho = 3H^2 + \frac{3k}{R^2} - \Lambda,$$

We write the above equation as,

$$\rho + \rho_{grav} + \rho_k + \rho_\Lambda \equiv \rho_{tot} = 0$$

where,

$\rho = \frac{Mc^2}{V}$ energy-density of matter.

$\rho_{grav} = -\frac{3H^2}{\kappa}$ negative energy density of the gravitational field

$\rho_k = -\frac{3k}{\kappa R^2}$ energy density of the tricurvature

$\rho_\Lambda = \frac{\Lambda}{\kappa}$ energy density of the vacuum

$V \propto R^3$ volume

We have added all energy-densities that could possibly exist, and the sum is the total zero-valued energy density. This is the reason why the total energy is also zero. We must take into consideration the negative energy and energy density of the gravitational field.As a consequence, there is no initial infinite energy density singularity.The total is zero.

7.13. The Cosmological Newtonian Limit of General Relativity

One usually goes from Newtonian gravity towards the generalization to GRT.Now, let us do the opposite trajectory, and find a Newtonian cosmological limit of Einstein´s theory.We already know, that, in the lowest limit,we find Poisson´s equation,

$$\frac{\partial^2 \Psi}{\partial x^\nu \partial x^\nu} = -4\pi G\rho$$

Whittaker defined a gravitational energy density, and I think I was one of the first to include the cosmological constant in it, so that, the real effective density in the above equation, should be ,

$$\rho \to \rho + 3p - 2\frac{\Lambda}{\kappa},$$

so we begin with the generalized Poisson´s,

$$\frac{\partial^2 \Psi}{\partial x^\nu \partial x^\nu} = -4\pi G(\rho + 3p - 2\frac{\Lambda}{\kappa})$$

Now, consider N distant stars distributed evenly at a distance R from the observer,and with masses M/N each one.Applying the above equation to each star, and summing all the N terms, we shall find, that, if each one contributes with a potential,

$$\phi = \frac{G}{RN}M$$

the total lhs of the generalized Poisson equations will become,

$$2GM/R^3 = 4\pi G(\rho + 3p - 2\frac{\Lambda}{\kappa})$$

A Machian type of solution could be,

$$\rho = \rho_0 R^{-2}$$

$$p = p_0 R^{-2}$$

$$\Lambda = \Lambda_0 R^{-2}$$

The above solution, has independent support in Chapter 8, so now let us assume this as a Machian distant stars limit of General Relativistic Cosmology.

We are now left with a disguised Brans-Dicke relation, to wit,

$GM/R = c^2\gamma \equiv 2\pi G(\rho_0 + 3p_0 - 2\frac{\Lambda_0}{\kappa})$
Look out that the last rhs needs to be positive for the Universe, i.e.

$$\rho_0 + 3p_0 - 2\frac{\Lambda_0}{\kappa} \geq 0$$

It is a positivity of energy requirement for the Universe.

Part IV

The Pioneers Anomaly

Chapter 8

The Pioneers Anomaly and a Machian Universe

8.1. Brief History of the Pioneers Anomaly

The Pioneers 10 and 11, were the first spacecrafts attempts of mankind to explore outer space.Their escaping orbits were hyperbolic,near the plane of the ecliptic, and they travel in opposite sides out of the Solar system.They departed by the years 1972 and 1973. In 2002, Pioneer 10, was about 12 billion km from the Earth.The power supplies of the Pioneer 11, stopped in 1995, when it was far 6.5 billion km from the Earth.Radiometric data, points to a constant deceleration of -9.10^{-8}cm.s^{-2}.This is in violation of Newton´s law of gravitation.This deceleration points towards the Sun or the Earth, it is unclear.

We shall examine in this book cosmological causes of this deceleration,like Universal rotation, or the varying speed of light, which are equivalent .Other explanations involve either a modification of Newton´s law,initially made by M.Milgrom (1983), on a heuristic basis, which was later shown to be a possible weak-field approximation of a gravitational tensor-vector and scalar theory by Bekenstein (2004) .On-board forces, like the thermal recoil of internally generated heat, were dismissed by Anderson et al. (1999), because they should be decreasing — not constant.They should also surge in elliptically bound orbits of other spacecraft, but it turns out that the Pioneers Anomaly only happens in hyperbolic motion.The secondary effect, that the space-probes are spinning down, is a proof of another cosmological effect, as we show in this book, related to the rotation of the Universe.

8.2. The Zero-Total Energy Machian Universe

We now shall propose a semi-Relativistic treatment of Mach's Principle, which means a zero-total energy Universe. (Feynman, 1962-1963; Berman, 2006; 2006a), have shown this meaning of Mach's Principle without considering a rotating Universe. We now extend the model, in order to include the spin of the Universe, and we replace Brans-Dicke traditional relation, $\frac{GM}{c^2R} \sim 1$, with three different relations, which we call the Brans-Dicke relations for gravitation, for the cosmological "constant" , and for the spin of the Universe.

We shall consider a "large" sphere, with mass M , radius R , spin L , and endowed with a cosmological term Λ , which causes the existence of an energy density $\frac{\Lambda}{\kappa}$, where $\kappa = \frac{8\pi G}{c^2}$. We now calculate the total energy E of this distribution:

$$E = E_i + E_g + E_L + E_\Lambda, \tag{8.2.1}$$

where $E_i = Mc^2$, stands for the inertial (Special Relativistic) energy; $E_g \cong -\frac{GM^2}{R}$ (the Newtonian gravitational potential self-energy); $E_L \cong \frac{L^2}{MR^2}$ the Newtonian rotational energy; and $E_\Lambda \cong \frac{\Lambda R^3}{6G}$ (the cosmological "constant" energy contained within the sphere).

If we impose that the total energy is equal to zero, i.e., $E = 0$, we obtain from (8.2.1):

$$\frac{GM}{c^2R} - \frac{L^2}{M^2c^2R^2} - \frac{\Lambda R^3}{6GMc^2} \cong 1. \tag{8.2.2}$$

As relation(8.2.2) above should be valid for the whole Universe, and not only for a specific instant of time, in the life of the Universe, and if this is not a coincidental relation, we can solve this equation by imposing that:

$$\frac{GM}{c^2R} = \gamma_G \sim 1\ , \tag{8.2.3}$$

$$\frac{L}{McR} = \gamma_L \sim 1, \tag{8.2.4}$$

and,

$$\frac{\Lambda R^3}{6GMc^2} = \gamma_\Lambda \sim 1, \tag{8.2.5}$$

subject to the condition,

$$\gamma_G - \gamma_L^2 - \gamma_\Lambda \sim 1, \tag{8.2.6}$$

where the $\gamma' s$ are constants having a near unity value.

We note for the future, that what matters is L^2, not L, so that a negative Lalso would solve the problem.

We now propose the following generalized Brans-Dicke relations, for gravitation, spin and cosmological "constant":

$$\frac{GM}{c^2R} = \gamma_G \sim 1\ , \tag{8.2.3}$$

$$\frac{GL}{c^3R^2} = \gamma_G \cdot \gamma_L \sim 1, \tag{8.2.7}$$

and,

$$\frac{\Lambda R^2}{6c^4} = \gamma_\Lambda \cdot \gamma_G \sim 1. \tag{8.2.8}$$

The reader should note that we have termed Λ as a "constant", but it is clear from the above, that in an expanding Universe, $\Lambda \propto R^{-2}$, so that Λ is a variable term. We also notice that $R \propto M$, and $L \propto R^2$.

The B.D. relation for spin, has been derived, on a heuristic procedure, which consists on the simple hypothesis that L should obey a similar relation as M (Sabbata and Sivaram, 1994). The first authors to propose the above R^{-2} dependence for Λ were Chen and Wu (1990), under the hypothesis that Λ should not depend on Planck's constant, because the cosmological "constant" is the Classical Physics response to otherwise Quantum effects that originated the initial energy of the vacuum. Berman, as well as Berman and Som, have examined, along with other authors, a time dependence for Λ(see for example, Berman, 1991; 1991a).

It must be remarked, that our proposed law (8.2.3), is a radical departure from the original Brans-Dicke (Brans and Dicke, 1961) relation, which was an approximate one, while our present hypothesis implies that $R \propto M$. With the present hypothesis, one can show that, independently of the particular gravitational theory taken as valid, the energy densities of the Machian Universe obey a R^{-2} dependence (see Berman, 2006; 2006a; Berman and Marinho, 2001).

Consider, for instance, the inertial energy density. We define,

$$\rho = \frac{Mc^2}{V}, \tag{8.2.9}$$

while,

$$V = \alpha R^3, \quad (\alpha = \text{constant}) \tag{8.2.10}$$

where ρ and V stand for energy density and tridimensional volume, we find:

$$\rho = \left[\frac{\gamma_G}{G\alpha}\right] R^{-2}. \tag{8.2.11}$$

For all other kinds of energy densitites (gravitational, lambda, or spin-originated), we also find the R^{-2} dependence, when we admit the above Brans-Dicke generalised relations, as the reader may easily check

If we apply the above relation, for Planck's and the present Universe, we find:

$$\frac{\rho}{\rho_{Pl}} = \left[\frac{R}{R_{Pl}}\right]^{-2}. \tag{8.2.12}$$

If we substitute the known values for Planck's quantities, while we take for the present Universe, $R \cong 10^{28}$ cm, we find a reasonable result for the present energy density. This shows that our result (relation 12.3.11), has to be given credit.

It should be remembered that the origin of Planck's quantities, say, for length, time, density and mass, were obtained by means of dimensional combinations among the constants for macrophysics (G for gravitation and c for electromagnetism) and for Quantum

Physics (Planck's constant $\frac{h}{2\pi}$). Analogously, if we would demand a dimensionally correct Planck's spin, obviously we would find,

$$L_{Pl} = \frac{h}{2\pi} \; . \tag{8.2.13}$$

From Brans-Dicke relation for spin, we now can obtain the present angular momentum of the Universe,

$$L = L_{Pl} \left[\frac{R}{R_{Pl}} \right]^2 \cong 10^{120} \left(\frac{h}{2\pi} \right) = 10^{93} \quad g\ cm^2\ s^{-1} \; . \tag{8.2.14}$$

This estimate was also made by Sabbata and Sivaram(1994), based on heuristic considerations(see also Sabbata and Gasperini, 1979).

If we employ, for the cosmological "constant" Planck's value, Λ_{Pl} ,

$$\Lambda_{Pl} \cong R_{Pl}^{-2} \; , \tag{8.2.15}$$

then, we shall find, in close agreement with the present value estimate for Λ (as found by recent supernovae observations), by means of the third Brans-Dicke relation:

$$\Lambda = \Lambda_{Pl} \left[\frac{R_{Pl}}{R} \right]^{-2} \; . \tag{8.2.16}$$

We refer to analysis in Section 12.6, in order to check that, corresponding the equation $E = 0$,which applies to the zero-total energy of the Universe, we also find $\bar{\rho} = 0$, i.e., the effective energy density of the Universe is also zero ($\bar{\rho} = \frac{dE}{dV} = 0$), where V represents volume. Thus, as these results are time-invariant, we assert that they are valid at time $t = 0$. Such being the case, we conclude that in the Machian Universe, there is no room for the so-called initial singularity (infinite total energy density at initial time).

APPLICATION 1 - THE PIONEERS ANOMALIES

Sabbata and Gasperini(1979), have calculated the angular speed of the Universe, for the present Universe. Though they mixed their calculations with some results obtained from Dirac's LNH (Large Number Hypothesis), including a time variation for the gravitational "constant", we now show that, if we take for granted that $G =$ constant, and by means of the generalized Brans-Dicke relations we find, by considering a rigid rotating Universe, whereby:

$$L = \pm M R^2 \omega, \tag{8.2.17}$$

so that,

$M\omega =$ constant , (because $L \propto R^2$ as we have shown earlier), we shall have:

$$\omega_{Pl} = \pm \frac{c}{R_{Pl}} = \pm 2 \text{ x } 10^{43} \;\; s^{-1} \; , \tag{8.2.18}$$

and, for the present,

$$\omega = \pm \tfrac{c}{R} \cong \pm 3 \text{ x } 10^{-18}\ s^{-1}. \tag{8.2.19}$$

It must be pointed out, that the result (8.2.18) seems to us that was not published elsewhere, up to now. This is, however, the first time that the above results are obtained by means of the zero-total energy hypothesis for the Universe. This is why we attribute this hypothesis to a Machian Universe; indeed, we believe that we can identify Mach's Principle, with this hypothesis.

Sabbata and Gasperini(1979), pointed out that the same numerical angular speed is obtained for Gödel's Universe, and also for the Sun's peculiar velocity through the cosmic microwave background.

We remark that $\gamma_G \cong 2$ is to be exact and not approximate, if we consider the result by Adler et al (1975), for the energy of a spherical mass, obtained by means of pseudotensors.

The Pioneers' anomaly, is described by a centripetal acceleration of an up to now unexplained nature, which affects two spaceships launched on opposite directions, which are by now in the outskirts of the Solar system. Its value is $a' \cong -9 \text{ x } 10^{-8} cm/\sec^2$.

For a Machian Universe, taken care of result (8.2.11),we can obtain the value for an ubiquitous centripetal acceleration,

$$a = -\omega^2 R\,. \tag{8.2.20}$$

If $R \cong 10^{28} cm$, as is known for the causally related Universe, we find:

$$a = -9 \text{ x } 10^{-8} cm/\sec^2 \cong a'. \tag{8.2.21}$$

It is necessary to point out that, for a Machian Universe, we should have this extra acceleration, along the direction pointing from the observed to the observer. It affects any two pairs of, observer versus observed, points in space. The striking match between a and a' must point to a possible solution to the Pioneers' anomaly; the only necessary hypothesis is that the Universe is endowed with the Machian properties shown above.

There is a secondary Pioneers Anomaly, reported by researchers. The two spacecrafts are loosing spin.

The first one has already been tackled by me. Now, let us face the second one. The spins of the Pioneers were telemetered, and as a surprise, it shows that the on-board measurements yield a decreasing angular speed, when the space-probes were not disturbed. Turyshev and Toth (2010), published the graphs (Figures 2.16 and 2.17 in their paper), from which it is clear that there is an angular deceleration of about 0.1 RPM per three years, or,

$$\alpha \approx -1.2 \times 10^{-10} \text{rad/s}^2.$$

As the diameter of the space-probes is about 10 meters, the linear acceleration is practically the Pioneers anomalous deceleration value ,in this case, -6.10^{-8} $cm.s^{-2}$.The present solution of the second anomaly, confirms our first anomaly explanation.

As the Universe expands, the cosmological component of the spacecraft´spin decreases.Of course, there may be other kinds of spin, but the cosmological effect is to decrease it.

APPLICATION 2 - MAGNETIC FIELD OF THE UNIVERSE

In two recent chapters of books (Berman, 2006; 2006a) has proposed a new interpretation for Brans-Dicke relation, which, instead of being an approximate relation only valid for the present Universe, should be interpreted as meaning that the mass M , of the causally related Universe, is directly proportional to the radius R , in the entire life of the Universe. In the same references, it is shown that, the new interpretation of Brans-Dicke relation, along with the hypothesis that the cosmological "constant" varies with R^{-2} , arise from the imposition that the total energy of the Universe, is zero-valued. Sabbata and Sivaram(1994), have shown that, in analogy with Brans-Dicke approximate relation, one could state another similar one for the spin of the Universe L . Again, Berman(2006a), extending his conjectures on the zero-total energy of the Universe, and including in the total energy, a term representing the rotational energy, derived Sabbata and Sivaram(1994) relation, as an exact formula, indicating that L varied with R^2 during all times. In all cases, Berman has made the hypothesis, that the fraction of each kind of energy participation, did not vary with time, when taken as fractions of Mc^2 .

We now extend Berman's hypotheses, while keeping a new term which contributes to the total energy of the Universe, dictated by the magnetic field. The fraction of magnetic energy participation to the total energy, is nevertheless kept in a 10^{-3} orders of magnitude, because we take for granted that the observed equipartition between the microwave background radiation, and magnetic field energies, for interstellar media, point out to a similar fraction in the magnetic field of the Universe. We impose that such fraction endures for the entire history of the Universe; in fact, this means that we adjust our Machian relation for the magnetic field, in order that its present value should be around 10^{-6} Gauss.

As the fractions of energy, of any kind, in the Machian Universe, according to our theory, are to be maintained, we take for granted, that any kind of energy's density, vary with R^{-2} , as has been shown, for the total energy density, by Berman(2006; 2006a) and Berman and Marinho Jr(2001). For each type of energy, we would have a constant fraction of the total energy, i.e., constant in time. For the inertial energy density, we would have:

$$\rho = \frac{Mc^2}{\frac{4}{3}\pi R^3}. \tag{8.2.22}$$

From Brans-Dicke relation, as modified by Berman, we have:

$$\frac{GM}{Rc^2} = \gamma = \text{ constant } \sim 1. \tag{8.2.23}$$

From (8.2.22) and (8.2.23), we obtain the desired dependence, $\rho \propto R^{-2}$.

The energy density associated with a magnetic field B is given by:

$$\rho_B = \frac{B^2}{8\pi}. \tag{8.2.24}$$

In mass units, we have to divide the second member of (8.2.24) by c^2 . The total energy fraction for the magnetic field, relative to Mc^2 would be given by:

$$\left[\frac{4}{3}\pi R^3\right]\left[\frac{B^2}{8\pi c^2}\right]\left[Mc^2\right]^{-1} = \gamma_B \cong 10^{-6}. \tag{8.2.25}$$

We then find that $B \propto R^{-1}$ because, in fact, from (8.2.25) we have:

$$B^2 = 12\ c^4\, \gamma\, \gamma_B\, G^{-1} R^{-2}. \tag{8.2.26}$$

We then find, for the present Universe, with $R \cong 10^{28}$ cm, $B \cong 10^{-6}$ Gauss .

For Planck's Universe, we would have, with $R_{Pl} \cong 10^{-33}$ cm,

$$B_{Pl} = B\left[\frac{R}{R_{Pl}}\right] \cong 10^{55} \text{ Gauss.}$$

This last value is larger than the maximum limit for the magnetic field not to provoke instabilities in the vacuum, according to a recent analysis made through Quantum Electrodynamics theory (QED), by Shabad and Usov(2006). That being the case, we can imagine this fact as causing the eruption of the inflationary phase, mediately after Planck's time.

We have derived the dependency of the magnetic field with R^{-1} , from the zero-total energy conjecture, in a Machian Universe, (see relation 8.2.25 above),obtaining a result valid during the lifespan of the Universe.

We remark that Sabbata and Sivaram(1994) obtained for the Planck's magnetic field, $B'_{pl} \sim 10^{58}$ Gauss, which is larger than in our estimate. But the law of variation of the magnetic field, with the age of the Universe, is our own.

APPLICATION 3 - TIME-VARYING NEUTRINO MASS

The subject of mass-varying neutrinos, has been very recently given attention (Horvat, 2005; Fardon et al, 2003; Kaplan, 2004). Berman (2007; 2007a) has made, with the help of F.M. Gomide, an historical review of the neutrino research and considered its rôle as dark matter in the Universe. It has been asserted, that 67% of the energy density of the

Universe, is due to a cosmological "constant" energy. The rest of the energy density is fractionated in two parts: 5% as visible mass and 28% as dark matter. Let us suppose that dark matter is constituted by neutrinos with non-zero rest mass. Berman(2006, 2006a, 2006b)has suggested that, if Mach's principle is understood as meaning that the total energy of the Universe is null, and if each particular energy contribution to the total energy density, has constant participation during the whole history of the Universe, one may obtain different

Machian relations. These Machian relations, of which, the Brans-Dicke (Brans and Dicke, 1961) relation is a particular case, should not, according to Berman, be viewed as just coincidental with the present Universe.

Suppose that the total energy is given by:

$$E = Mc^2 - \frac{GM^2}{2R} + 4\pi\Lambda\frac{R^3}{3\kappa} + \frac{L^2}{MR^2} \ , \tag{8.2.27}$$

where the four terms to the right of relation (8.2.27) represent respectively the inertial, gravitational, cosmological constant's and rotational energies.

When we impose,

$$E = 0 \ , \tag{8.2.28}$$

we obtain the generalised Brans-Dicke relations of above, which are: (8.2.3), (8.2.4), (8.2.5) and (8.2.6). Then, we can check that all energy densities are proportional to R^{-2} .

We can also write:

$$\rho_{tot} \equiv \frac{3E}{4\pi R^3} \equiv \rho + \rho_{grav} + \rho_\Lambda + \rho_{rot} = 0$$

where,

$$\rho = \frac{3Mc^2}{4\pi R^3} \qquad \text{energy density of matter}$$

$$\rho_{grav} = -\frac{3GM^2}{8\pi R^4} \qquad \text{negative energy density of the field}$$

$$\rho_\Lambda = \frac{\Lambda}{\kappa} \qquad \text{vacuum}$$

$$\rho_{rot} = \frac{3L^2}{4\pi MR^5} \ , \qquad \text{rotational energy density} \tag{8.2.29}$$

[where ρ_{grav} is the (negative) gravitational self-energy density correspondent to the self-gravitational energy, ($E_{grav} \cong -\frac{GM^2}{2R}$) which is representative of the total energy of the Schwarzschild's metric - see relation (8.2.27)].

In the spirit of inflationary Cosmology (Guth, 1981), we identify, for the present Universe, the "total" energy density with the critical density, so that we would have:

$$\rho_{crit} \cong 2 \text{ x } 10^{-29} \text{ g / cm}^3 = -\rho_{grav}.$$

In the next few paragraphs, we estimate neutrinos average mass, and its time variation. But, we observe that, if dark matter is a fraction of ρ_{TOT} , this fraction will also depend on R^{-2} , so as to keep all relative components equally balanced along time.

A theory for neutrinos energy density

As we have noticed before the energy density of dark matter, to be identified with neutrinos, shall be given by:

$$\rho_\nu = 0.27\rho_{crit}. \qquad (8.2.30)$$

Berman (2006c) along with others (see Sabbata and Sivaram, 1994) have estimated that the Universe possess a magnetic field which, for Planck's Universe, was as huge as 10^{55} Gauss. The relic magnetic field of the present Universe is estimated in 10^{-6} Gauss. We can then, suppose that all neutrinos' spins have been aligned with the magnetic field. On the other hand, the spin of the Universe is believed to have increased in accordance with Machian relations(8.2.3) to (8.2.8), which entails that $L \propto R^2$. If we call n the number of neutrinos in the present Universe, and n_{Pl} its value for Planck's Universe, we may write:

$$\frac{n}{n_{Pl}} = \frac{L}{L_{Pl}} = 10^{120}. \qquad (8.2.31)$$

Then,

$$n = n_{Pl}\left[\frac{R}{R_{Pl}}\right]^2. \qquad (8.2.32)$$

We have just obtained the relation for the increase of the number of neutrinos with R^2

.

Now, we write the energy density of neutrinos,

$$\rho_\nu \cong \frac{n m_\nu c^2}{\frac{4}{3}\pi R^3}, \qquad (8.2.33)$$

where m_ν is the rest mass of the average neutrino.

If we impose relation (8.2.30) and simultaneously, relations (8.2.29) and (8.2.33), we conclude two things:

$$1^{st}.) \quad \rho_\nu = 0.27\rho_{Pl}\left[\frac{R}{R_{Pl}}\right]^{-2}. \qquad (8.2.34)$$

$$2^{nd}.) \quad m_\nu = \frac{\rho_{Pl} R_{Pl}^4}{R}.$$

We see now that while the number of neutrinos in the Universe increases with R^2 , the rest mass decreases with R^{-1} ; we may obtain, with $R \cong 10^{28}$ cm, that the rest mass of neutrinos should be, in the present Universe:

$$m_\nu \cong 10^{-65} \text{ g}. \qquad (8.2.35)$$

F.M. Gomide (1963), has estimated the mass of neutrinos a long time ago, finding, in a seminal paper , the value, 10^{-65} g. One of the two different arguments by Gomide (Gomide, 1963), was in fact that he equated $m_\nu c^2$ with the self-gravitational energy of the proton.

A law of variation for the number of neutrinos in the Universe has been found. A law of variation for the rest mass of neutrinos was also found.

We remind the reader that Kaluza-Klein's cosmology (Wesson, 1999; 2006; Berman and Som, 1993), consider time varying rest masses, in a penta-dimensional space-time-matter, of which the fifth coordinate is rest mass. The above results can not be rejected, for the time being, by any known data. We point out, that some of the features of the present calculation, are originated from a seminal paper by Sabbata and Gasperini (1979).

APPLICATION 4 - ARE MASS AND LENGTH QUANTIZED?

After having derived the generalized Brans-Dicke relations, as obtained earlier by Berman, and the possibility of time-varying neutrinos rest-mass (Berman, 2007; 2007a), when the Universe is Machian, we now introduce the definitions of micromass and macromass, as well as those of microlength and macrolength, in the spirit of Wesson's suggestions (Wesson, 2006). We show that by obtaining such quantities for Planck's time, and the present Universe, a) both "micros" coincide with Planck's mass and length, while for the present Universe, macrolength stands as the radius of the causal Universe, while macromass represents the mass of the Universe. However, we find a quantum of mass ("*gomidium*"), which we associated earlier with neutrino's average rest-mass, and a quantum of length ("*somium*"), to which we suggest interpretations.

Macromass and micromass

Wesson (2006), by citing Desloge(1984), comments that by means of the four fundamental "constants", Planck's (h), Newton's (G), speed of light (c), and cosmological (Λ), one can obtain two different kind of mass, the micromass (m), and the macromass (M), given by:

$$m = \left(\frac{h}{c}\right)\Lambda^{1/2}, \tag{8.2.36}$$

and,

$$M = \frac{c^2}{G}\Lambda^{-1/2}\,. \tag{8.2.37}$$

Notice that the above constant tetrad, is, of course, overlapping. Nevertheless, Wesson dealt only with the present values for the cosmological "constant", $\Lambda = \Lambda_U \approx 10^{-56}\ \text{cm}^{-2}$, and then he found,

$$m_{(U)} \approx 10^{-65}\ \text{g}\,, \tag{8.2.38}$$

and,

$$M_{(U)} \approx 10^{56}\ \text{g}\,. \tag{8.2.39}$$

We call the present Universe's micromass ($m_{(U)}$), as the present value for neutrinos' average mass, and we shall see that it represents a mass-quantum, i.e., the minimum mass in the present Universe. On the other hand, the present Universe's macromass ($M_{(U)}$), is approximately the mass of the present Universe (M_U) .

What Wesson overlooked, is that, when we apply the definitions (8.2.36) and (8.2.37), by plugging, $\Lambda = \Lambda_{PL} \approx L_{PL}^{-2} \approx 10^{-66}$ cm^{-2} , which stands for Planck's time values, we find that micromass and macromass coincide approximately with Planck's mass,

$$M_{(PL)} = m_{(PL)} = M_{PL} \approx 10^{-5} \text{ g} . \tag{8.2.40}$$

We are led to consider that, macromass, is always associated to the mass of the Universe (M_U), either in the very early Universe or in the present one.

As to the micromass, we can see that it coincides with the previously estimated neutrinos' average mass, either for the present Universe or for the very early one. We baptize this mass as the quantum mass value (*gomidium,* after F.M.Gomide): it is a time-varying mass, as we have shown before.

Quantization of geometry

We now show, that associated with micromass and macromass, we have two distinct length values, which come associated to the present and Planck's Universe.

For each mass, we associate two kinds of lengths, namely, the macrolength, (λ_ν), and the microlength (l_ν); the first one, is Compton's wavelength, given by,

$$\lambda_\nu = \frac{\hbar}{mc} . \tag{8.2.41}$$

The microlength, is a gravitationally associated length with mass, which we term the quantum of length, or *somium* , when we apply the micromass,

$$l_\nu = \frac{Gm}{c^2} . \tag{8.2.42}$$

By plugging numerical values, we find, a microlength, l_ν ,for the present Universe, with $m = m_\nu$, that,

$$l_{\nu(U)} \approx 10^{-91} \text{ cm}, \tag{8.2.43}$$

while, the macrolength is given by,

$$\lambda_{\nu(U)} \approx 10^{28} \text{ cm} . \tag{8.2.44}$$

On the other hand, for Planck's Universe,

$$\lambda_{\nu(PL)} \approx 10^{-33} \text{ cm} , \tag{8.2.45}$$

and,

$$l_{\nu(PL)} \approx 10^{-33} \text{ cm} . \tag{8.2.46}$$

One can check, that the microlength for the present Universe represents a quantum, which we call the present Universe's value for the *somium* . Macrolength is represented by the radius of the Universe.

For Planck's Universe, the *somium* coincides with Planck's length; both macro and micro, then coincide with the Planck's radius.

We have found that the micromass and microlength represent quanta of mass and length. We call them, respectively, *gomidium* and *somium*, but their numerical values are time-varying: present day's *gomidium* is 10^{-65} g, while *somium* is about 10^{-91} cm. Planck's values for *gomidium* and *somium*, are respectively given by Planck's mass and Planck's length. We have thus hinted that mass is quantized, but geometry is altogether. As gravitation is associated with geometry, quantization of the latter, implies on the former: it seems that quantum gravity has been found.

Removal of initial singularity in Cosmology

From what has been dealt before, in this Section, it follows that, the initial singularity of the Universe, is now deleted from our Machian Universe picture: as the total energy density is zero, and the total energy of the Universe $E = 0$, we have not to deal with infinities while calculating those quantities in the limit $R \to 0$.

8.3. Dirac's LNH with Time-Varying Fundamental "Constants"

We shall study a generalisation of Dirac's LNH Universe, with the introduction of time-varying speed of light, which causes a time-varying fine-structure "constant", and a possible rotation of the Universe, either for the present time, or for inflationary periods. This Section is a sequel to a previous paper dealing with the more or less equivalent consequences of a time-varying electric and magnetic permittivity(Berman,2009).

The rotation of the Universe (de Sabbata and Sivaram, 1994; de Sabbata and Gasperini, 1979) may have been detected experimentally by NASA scientists who tracked the Pioneer probes, finding an anomalous deceleration that affected the spaceships during the thirty years that they took to leave the Solar system. This acceleration can be explained through the rotation of the Machian Universe (Berman, 2007b). A universal spin has been considered by Berman (2008b; 2008c).

A time-varying gravitational constant, as well as others, were conceived by P.A.M. Dirac (1938; 1974), Eddington (1933; 1935; 1939), Barrow (1990) through his Large Number Hypothesis. Later, Berman supplied the GLNH – Generalised Large Number Hypothesis (Berman, 1992; 1992a; 1994). This hypothesis arose from the fact that certain relationships among physical quantities, revealed extraordinary large numbers of the order 10^{40} . Such numbers, instead of being coincidental and far from usual values, were attributed to time-varying quantities, related to the growing number of nucleons in the Universe. In fact, such number N , for the present Universe, is estimated as $\left(10^{40}\right)^2$. The number is "large" because the Universe is "old". At least, this was and still is the best explanation at our disposal.

The four relations below, represent respectively, the ratios among the scalar length of the causally related Universe, and the Classical electronic radius; the ratio between the electrostatic and gravitational forces between a proton and an electron; the mass of the Universe divided by the mass of a proton or a nucleon; and a relation involving the cosmological constant and the masses of neutron and electron.

If we call Hubble's constant H ; electron's charge and mass e , m_e ; proton's mass m_p , cosmological constant Λ , speed of light c , and Planck's constant h , we have:

$$\frac{cH^{-1}}{\left(\frac{e^2}{m_e c^2}\right)} \cong \sqrt{N} \ . \qquad (8.3.1)$$

$$\frac{e^2}{Gm_p m_e} \cong \sqrt{N} \ . \qquad (8.3.2)$$

$$\frac{\rho(cH^{-1})^3}{m_p} \cong N \ . \qquad (8.3.3)$$

$$ch(m_p m_e/\Lambda)^{1/2} \cong \sqrt{N} \ . \qquad (8.3.4)$$

We may in general have time-varying speed of light $c = c(t)$; of $\Lambda = \Lambda(t)$; of $G = G(t)$; etc. We define the fine structure "constant" as,

$$\alpha \equiv \frac{e^2}{\hbar\, c(t)} \ , \qquad (8.3.5)$$

and consider $\alpha = \alpha(t)$, because of the time-varying speed of light.

POWER-LAW VARIATIONS

One can ask whether the previous Section's constant-variations could be caused by a time-varying speed of light: $c = c(t)$. We refer to Berman (2007) for information on the experimental time variability of α . Gomide (1976) has studied $c(t)$ and α in such a case, which was later revived by Barrow (1998 ; 1998a ; 1997); Barrow and Magueijo (1999); Albrecht and Magueijo (1998); Bekenstein (1992). This could explain also Supernovae observations. We refer to their papers for further information. Our framework now will be an estimate made through Berman's GLNH.

We express now Webb et al's (1999; 2001) experimental result as:

$$\left(\frac{\dot{\alpha}}{\alpha}\right)_{\text{exp}} \simeq -1.1 \text{ x } 10^{-5}\, t^{-1} \ , \qquad (8.3.6)$$

where t responds for the age of the Universe.

From (8.3.5) we find:

$$\frac{\dot{\alpha}}{\alpha} = -\frac{\dot{c}}{c} \ . \tag{8.3.7}$$

Again, we suppose that the speed of light varies with a power law of time:

$$c = At^n \qquad (A = \text{constant}) \ . \tag{8.3.8}$$

From the above experimental value we find:

$$n \approx 10^{-5} \ . \tag{8.3.9}$$

From (8.3.8) and (8.3.9) taken care of (8.3.7), we find:

$$\frac{\dot{\alpha}}{\alpha} = -\frac{\dot{c}}{c} = nt^{-1} \ . \tag{8.3.10}$$

From relations (1), (2), (3) and (4) we find:

$$N \propto t^{2+6n} . \tag{8.3.11}$$

$$G \propto t^{-1-3n} . \tag{8.3.12}$$

$$\Lambda \propto t^{-2-4n} \ . \tag{8.3.13}$$

$$\rho \propto t^{-1+3n} \ . \tag{8.3.14}$$

We see that the speed of light varies slowly with the age of the Universe. For the numerical value (8.3.9), we would obtain:

$$N \propto t^{2.0001} \ , \tag{8.3.15 }$$

and then:

$$G \propto t^{-1.00005} \ . \tag{8.3.16}$$

$$\Lambda \propto t^{-2.0001} \ . \tag{8.3.17}$$

$$\rho \propto t^{-0.99995} \ . \tag{8.3.18}$$

This is our solution, based on Berman's GLNH, itself based on Dirac's work (Dirac, 1938; 1974). A pre-print with a preliminary but incomplete solution was already prepared by Berman and Trevisan (2001; 2001a; 2001b).

As a bonus we found possible laws of variation for N, G, ρ, and Λ. The Λ–term time variation is also very close and even, practically indistinguishable, from the law of variation $\Lambda \propto t^{-2}$.

It is clear that in this Section's model, the electric permittivity of the vacuum, along with its magnetic permeability, and also Planck's constant are really constant here. We point out again, that in the long run, it will be only when a Superunification theory becomes available, that the different models offered in the literature, could be discarded, (hopefully) but one.

EXPONENTIAL INFLATION

On remembering that relations (8.3.1) and (8.3.3) carry the radius of the causally related Universe, cH^{-1} , we substitute it by the exponential relation,

$$R = R_0 e^{Ht}. \tag{8.3.19}$$

With the same arguments above, but, substituting, (8.3.8) by the following one,

$$c = c_0 e^{\gamma t} \quad , \quad (c_0 \, , \, \gamma \, = \, \text{constants}) \tag{8.3.20}$$

we would find:

$$N \propto e^{[H+2\gamma]\, t}, \tag{8.3.21}$$

$$G \propto e^{-\left[\frac{H}{2}+\gamma\right]\, t}, \tag{8.3.22}$$

$$\rho \propto e^{-2[H-\gamma]\, t}, \tag{8.3.22a}$$

and,

$$\Lambda \propto e^{-H\, t}. \tag{8.3.22b}$$

It seems reasonable that inflation decreases the energy density, and the cosmological term while N grows exponentially; of course, we take $H > \gamma$.

ROTATION OF THE UNIVERSE

A closely related issue is the possibility of a Universal spin. Consider the Newtonian definition of angular momentum L ,

$$L = RMv, \tag{8.3.23}$$

where, R and M stand for the scale-factor and mass of the Universe.

For Planck's Universe, the obvious dimensional combination of the constants $\bar{h}$, c , and G is,

$$L_{Pl} = \pm\,\bar{h}. \tag{8.3.24}$$

From (8.3.23) and (8.3.24), we see that Planck's Universe spin takes a speed $v = c$. For any other time, we take, then, the spin of the Universe as given by

$$L = \pm RMc\,. \tag{8.3.25}$$

In the first place, we take the known values of the present Universe:

$$R \approx 10^{28} cm\,,$$

and,

$$M \approx 10^{55} grams\,,$$

so that,

$$L = \pm 10^{93} cm.gram.cm/s = 10^{120}\,\bar{h}\,. \tag{8.3.26}$$

We have thus, another large number,

$$\frac{L}{\bar{h}} \propto N^{3/2}\,. \tag{8.3.27}$$

For instance, for the power law, as in standard cosmology, we would have ,

$$L \propto t^{3+9n} = t^{3(1+3n)}\,. \tag{8.3.28}$$

For exponential inflation,

$$L \propto e^{\frac{3}{2}[H+2\gamma]\,t}. \tag{8.3.29}$$

We now may guess a possible angular speed of the Universe, on the basis of Dirac's LNH. For Planck's Universe, the obvious angular speed would be:

$$\omega_{Pl} = \pm\frac{c}{R_{Pl}} \approx 2 \text{ x } 10^{43} s^{-1}, \tag{8.3.30}$$

because Planck's Universe is composed of dimensional combinations of the fundamental constants. I recall a paper by Arbab (2004), that attaches a meaning to the above angular speeds, as yielding minimal accelerations in the Universe. The argument runs as follows. From manipulation with the constants that represent the Universe (c, h, G) we can construct, not only Planck´s usual quantities, but also a dimensionally correct acceleration. With this acceleration, we would construct, if we call it a centripetal $a = -\omega^2 R$ term, the angular speed of our present calculation. But Arbab failed to interpret the existing

Planck´s constant as representing an angular rotation. However, he says that this centripetal acceleration is a consequence of the vacuum energy, and calculates correctly its present value.

In order to get a time-varying function for the angular speed, we recall Newtonian angular momentum formula,

$$L = R^2 M \omega \,. \tag{8.3.31}$$

In the case of power-law c – variation , we have found, from relation (8.3.27), that, $L \propto N^{3/2}$, but we also saw from (8.3.31) that $L \propto \rho R^5 \omega$, because $R = cH^{-1} \propto \sqrt{N}$ and

$M \propto \rho R^3 \propto N$.

Then, we find that,

$$\omega = \omega_0 t^{-1+6n} = AR^{-(1-6n)} \qquad (\ \omega_0 \ , A = \text{constants}\ ) \,. \tag{8.3.31a}$$

We are led to admit the following relation:

$$|\omega| \lessapprox \tfrac{c}{R} . \tag{8.3.32}$$

For the present Universe, we shall find,

$$|\omega| \lessapprox 3 \text{ x } 10^{-18} s^{-1} \,. \tag{8.3.33}$$

It can be seen that present angular speed is too small to be detected by present technology.

For the inflationary model, we carry a similar procedure:

$$\omega \propto \frac{N^{\frac{3}{2}}}{R^5 \rho} = \pm e^{\left[-\frac{9}{2}H+\gamma\right]t} \,. \tag{8.3.34}$$

The condition for a decreasing angular speed in the inflationary period, is, then,

$$\gamma < \tfrac{9}{2} H \,. \tag{8.3.35}$$

PROS AND CONS OF THE PRESENT CALCULATIONS

Critical appraisals of the above calculations, center on the four following arguments:

I - do the time variations, $G(t)$, $\rho(t)$, and $\Lambda(t)$ proposed above, violate Einstein's field equations?

II - if Einstein's theory does not apply, which one does? And then, do the new equations reduce to Einstein's in a proper limit?

III - if there is rotation, would it not imply some preferred direction in the Universe?

IV - can order of magnitude calculations, be valid in order to get insights on the Universe?

We now reply:

1st) Dirac never proposed LNH as part of GRT (General Relativity Theory), neither do I.

2nd) Dirac's LNH is a foil for testing hypotheses, like the theoretical frameworks of scalar-tensor cosmologies with lambda (see for instance Berman, 2007a). In such theories, our present results may be included (Berman, 2007a).

3rd) According to the Machian approach by Berman (Berman, 2007b; 2008a), the kind of rotation to be expected in the Universe, has no unique axis of rotation; we know that there is a Machian rotation, because each "observer" sees any "observed", far away (at cosmological distances), with the centripetal acceleration that identifies the "Machian rotation". It is a rotation, like in the Gödel Universe (Adler et al., 1975).

4th) Dirac's Universe, though appealing, does not stand as a mathematically correct solution of any gravitational theory, like for instance, General Relativity. It is more of a tool, that identifies possible physical effects in the Universe.

We hope to have clarified the former cons, with the latter pros.

8.4. On Sciama's Machian Origin of Inertia

The advancement of scientific knowledge has produced, in the aftermath of Newtonian theory, the surge of General Relativity and alternative versions of it. On a different line of thought, Sciama[1] has put forward a gravity theory based on an electrodynamical analogy.

The research on the origin of Inertia is a problem that involved passionately a number of physicists especially from the crisis of Classical Physics and from the birth of General Relativity. Einstein himself underlined the importance of his meditation on this argument in the development of his theory of gravitation. Moreover he formulated precisely his reflections on inertia origin in the Mach's Principle - that in its simpler form says that inertia properties of matter are determined in some manner by the other bodies of the Universe. Even though with the development of General Relativity Einstein rejected explicitly his first considerations on Inertia, the Principle represented, in a compact form, a research project that guided many scientists in the development of gravitational theories alternative to General Relativity (for example Brans-Dike Theory) - with the following cosmological implications - and the reformulations of Classical Mechanics based on Mach's reflections on Inertia (see for example the model proposed by Shrödinger).

This short introduction has only the simple task to set in his historical background the work submitted by the present author, that moves on these lines of thoughts. This letter, has a double structure: in the beginning part we present changes in the Sciama cosmological model (linked to gravitational theories alternative to General Relativity) with the introduction of the expansion and rotation of the Universe, and in the second part of the article we use the results just obtained (the angular velocity of the Universe) in the calculation of a constant of Graneau and Graneau theory. Moreover at the end of this letter, I present in a simple form an argument proposed by Berry, where the Machian analysis of inertia and the alternative cosmological model (like Sciama and Brans-Dike model) shows some common aspects.

The suggestion of a Universe in rotation - certainly the most interesting and the most problematic - is better understood in my prior papers ([3]-[8]), so leaving a small gap in the presentation. Also the Sciama cosmological model is presented in a manner that I myself consider a bit concise.

Another critical point, a key argument to link Machian Universe to Graneau and Graneau Theory, is the extremely concise - but I think extremely interesting - presentation of my interpretation of the Mach Principle as the mathematical representation of the zero-total-energy of the Universe and the consequent description of the Pioneer effect [3] - [8].

This observation also creates interesting links among research programs, like the cosmological model based on Field Theory and Newtonian Machian Physics, that are often considered in competition.

I consider this subject as an original contribution to the debate on the origin of inertia and the connected argument of the cosmological model: so I think the arguments proposed in this manuscript are very suitable.

It is shown bellow, that Sciama's paper can be generalised, by including, in the same model, a rotating and expanding Universe; this also implies that the total energy density of the Universe is null, at each point of space; on summing for the whole Universe, we obtain zero-total energy. A possible angular speed proportional to the inverse of the scale-factor is found, which is of the same form of that found by Berman(2007; 2007a; 2007b; 2008; 2008a; 2008b).

Then, we review a theory with the otherwise "arbitrary" constant B in the inertia formula, by Graneau and Graneau[2]. The numerical value was in fact, found by us, in accordance with a possible rotation of the Universe. In another Section, a Machian argument by Berry[9] is shown to be related with Sciama's theory.

SCIAMA'S INERTIA MODEL

In order to fulfill a theoretical need for accounting the inertia properties of matter, Sciama supposes that gravitation is analogous to electrodynamics. The working hypothesis is that inertial and gravitational forces cancel each other at any point of space, so that the total field is null. We combine first the calculations of expansion and rotation of the Universe; each one was treated by Sciama, isolated.

For a rest particle, the "electric" potential contribution from the whole Universe, as observed in time "t" is given by:

$$\Phi = -\int \frac{\rho}{r} dV. \tag{8.4.1}$$

If the density is uniform, we have from (8.4.1):

$$\Phi \cong -2\pi\rho c^2\tau^2 , \tag{8.4.2}$$

while, by symmetry, the "vector" potential is null:

$$\vec{A} = 0. \tag{8.4.3}$$

In formula (8.4.2), τ is associated with H^{-1}, where H stands for Hubble's parameter.

If the particle moves with a linear velocity $\vec{v}$, and we add Hubble's expansion, its total velocity is $[\vec{v}+\vec{r}H]$, at any point in the Universe. In the first approximation, Φ keeps approximately the same form as above. However, the vector potential $\vec{A}$, when $\vec{v}$ does not depend on $\vec{r}$, will be given by:

$$\vec{A} \cong -\int \frac{\rho\vec{v}}{cr} dV \cong \frac{\Phi\vec{v}(t)}{c} . \tag{8.4.4}$$

The "electric" field is given by the usual electromagnetic formula:

$$\vec{E} = -\nabla\Phi - \frac{1}{c}\frac{\partial\vec{A}}{\partial t} \cong -\frac{1}{c^2}\Phi\frac{\partial\vec{v}}{\partial t}. \tag{8.4.5}$$

The above is called in this case, a gravitoelectric field.

On the other hand, we would have a null gravitomagnetic field,

$$\vec{B} = \nabla \times \vec{A} \cong 0 . \tag{8.4.6}$$

Now, let us have a body of mass M , placed in the Universe. In the rest frame of a particle at rest, the total "electric" field will be given by:

$$E_{TOT} \cong -\frac{1}{c^2}\Phi\frac{d\vec{v}}{dt} + \left[\frac{M}{r^2} + \frac{\phi}{c^2}\frac{d\vec{v}}{dt}\right]. \tag{8.4.7}$$

In the above, ϕ is the usual potential of the body with mass M , on a test particle, i.e.

$$\phi = -\frac{M}{r}. \tag{8.4.8}$$

Now, we write the total field as:

$$E_{TOT} \cong \frac{M}{r^2} + \frac{1}{c^2}\left[\phi - \Phi\right]\vec{a}, \tag{8.4.9}$$

where $\vec{a}$ is the total acceleration (of the Universe plus the body), relative to the test particle. Put it in other frame: the particle accelerates towards the rest body, relative to the whole Universe.

As this theory is of the electromagnetic type, it is a linear one. We superimpose, then, the cases of radial expanding, and rotating pictures.

In the first place, consider a non-rotating Universe; now we set a reference frame with origin at the body; in relativistic units, near the origin, we shall have the scalar and vector potentials given by:

$$\vec{A} = 0, \tag{8.4.10}$$

and,

$$\Phi \cong -1 \,. \tag{8.4.11}$$

In a rotating Universe, however, if the axis of rotation is the Z , then near the origin, we shall have:

$$A_x \cong \omega y,$$

$$A_y \cong -\omega x \,, \tag{8.4.12}$$

$$A_z \cong 0,$$

$$\Phi^{rot} \cong -\left[1+\omega^2 r^2\right]^{1/2} .$$

In this case, the total field, will be given by:

$$\vec{E}^{rot}_{TOT} = -\nabla\Phi^{rot} - \frac{\partial\vec{A}}{\partial t} \cong -\frac{\omega^2 r}{\left[1+\omega^2 r^2\right]^{1/2}} + \frac{M}{r^2} \,. \tag{8.4.13}$$

Now, we equate, on the test particle, inertia to gravitation, one balancing the other, so that the total field is zero, where this total is not the above only but added to $\vec{E}_{TOT}$, which equate to zero:

$$-\frac{\omega^2 r}{\left[1+\omega^2 r^2\right]^{1/2}} + \frac{2M}{r^2} - 2\pi\rho H^{-2} \cong 0 \,. \tag{8.4.14}$$

A particular solution, is composed by the two equalities below, whose sum retrieves the above one:

$$\frac{M}{r^2} \cong 2\pi\rho H^{-2} a \,. \tag{8.4.15}$$

(However, the Newtonian acceleration, in the above, is given by $a = -\frac{M}{r^2}$). $\frac{\omega^2 r}{\left[1+\omega^2 r^2\right]^{1/2}} \cong \frac{M}{r^2}$. (8.4.16)

Equality (15) is satisfied by the Whitrow-Randall expression,

$$G\rho H^{-2} \cong 1 \,. \tag{8.4.17}$$

The above is equivalent, because $\rho \cong \frac{M}{\frac{4}{3}\pi r^3}$, to the Brans-Dicke form,

$$\frac{GM}{r} = \gamma \sim 1 \ . \tag{8.4.18}$$

Notice that with the above, mass and radius are directly proportional with each other. If we now go to (8.4.16), and solve it with the approximation (8.4.18), and we take the angular speed, as given by the Machian expression,

$$\omega \cong \pm \frac{\alpha}{r} \qquad , (\alpha \leqslant c = 1), \tag{8.4.19}$$

where α is a constant that must be found later, and then, the total spin of the rotating Universe will be proportional to r^2 , i.e.:

$$L \cong Mr^2\omega = \gamma\alpha r^2 \propto r^2. \tag{8.4.20}$$

We see, that from (16), after some algebra, if $\gamma \cong 2$, then $\alpha \approx c$, and thus, we make contact with Berman models [3]-[8].

We notice that the local effect at each point of space, makes the total energy density equal to zero; it is given, in the electromagnetic case, by a Poynting-like electric field density,

$$\rho_{TOT} = E^2_{TOT}/8\pi \cong 0 \ . \tag{8.4.22}$$

When we sum for all points of space, we obviously find that the total energy of the Universe is zero-valued.

GRANEAU AND GRANEAU'S THEORY

Graneau and Graneau[2], discuss a version of inertia theory, that would be originated in a Machian Newtonian theory. According to those authors, it is an instantaneous action-at-a-distance theory. The consequence of such theory, is the "new" force law for inertia,

$$\Delta F_i = -\frac{a}{\pi^2 B}\left[\frac{m_0 m_x}{r^2}\right], \tag{8.4.23}$$

where, ΔF_i is the inertial force between two particles with masses m_0 and m_x , separated by a radial distance r , while B is a universal constant, relating ΔF_i with a universal relative acceleration a between the two objects.

As I have shown elsewhere, Mach's principle can be thought of, as the mathematical representation of the zero-total-energy of the Machian Universe. In another paper, Berman[4] calculated the existence of an anomalous universal acceleration which acts relative to each pair of "observer" and "observed" objects, due to the same Machian principle. The numerical value of the expected relative acceleration, acting in a radial direction from the "observed" to the "observer", is about $9 \times 10^{-8} cm/\sec^2$. This result was shown by Berman to agree and explain the so-called Pioneers' anomalous acceleration relative to the Earth, which is affecting both spaceships that travel in the outskirts of the Solar system, in two opposite directions relative to the Earth or approximately towards the Sun.

Graneau and Graneau[2], have failed to determine the numerical value of their constant B . By comparing with Newtonian law of gravitation, we may write, on the assumption that the acceleration to be met is the Pioneers' anomalous one,

$$\Delta F_i = -\frac{a}{\pi^2 B}\left[\frac{m_0 m_x}{r^2}\right] = -G\left[\frac{m_0 m_x}{r^2}\right], \qquad (8.4.24)$$

so that,

$$a = \pi^2 BG\,, \qquad (8.4.25)$$

where, G stands for Newton's gravitational constant.

When the numerical values above are plugged, we find:

$$B \cong 1.0\ m^{-2} kg^{-1}\,. \qquad (8.4.26)$$

This constant is the result of interactions B_i from all other masses in the Universe, in the treatment of Graneau and Graneau, where the B_i 's are determined by the mass of each "cause" , i.e., each mass in the Universe, divided by its distance to the given local point of space where they cause the inertia force.

Though we have found the otherwise undetermined numerical value for B , we point out that we do not agree with Graneau and Graneau, when they discard General Relativity in favor of Newtonian gravitation. In fact, it has been shown earlier, that cosmological models obeying Brans-Dicke-Whitrow-Randall-Sciama relation,

$$\frac{GM}{c^2 R} \cong 1\,, \qquad (8.4.26)$$

which is derived from the Machian zero-total-energy hypothesis, and which models are based on General Relativity theory or alternative generalizations of such theory, should be regarded as fulfilling the Machian property.

Conclusions

Sciama's linear theory, was generalised by including in the same model a rotating and at the same time expanding Universe. The resultant equations, are equivalent to the Machian treatment by Berman[3]-[8].

We have also found the constant B numerical value, but, nevertheless, we guess that we need not exclude other theories in favor of Newton's one. Another Section shows the deep value of Machian ideas. Indeed, inertia has been treated as Machian-originated.

8.5. Sciama's Machian Universe

Amidst several alternative theories of Gravity, which modify General Relativity, there stands an electrodynamical-type gravitational theory, put forward by Sciama (1953). The idea behind inertia, would be, according to Sciama, that the Universe obeys Mach's principle, i.e., at each point of space, the total force acting on a particle, is null, being composed by the second Newtonian law of force $m\vec{a}$, summed with a negative equal inertia force, which would be originated from the rest of the Universe and applied at that point. Thus, the total energy of the Universe would also be zero. The interaction of the Universe in a local point, would be made through linearized equations originated from Maxwellian-type fields (Arbab, 2004). Berman (2008d), has detailed the cosmological consequences of Sciama's theory, and showed that the Universe has expansion plus rotation, while the angular speed is inversely proportional to the scale-factor (the "radial" coordinate), and the Universe obeys Brans-Dicke relation in the form of the equality,

$$G\frac{M}{c^2R} = \gamma \sim 1 \ , \qquad (8.5.1)$$

where $\gamma =$ constant.

The purpose of the present, is to show that the Machian Universe, as described by Berman (2007, 2007a, 2007b, 2008, 2008a, 2008b), matches Sciama's linearized theory of "electrodynamical" gravitation; that (gravitational) radiation yields the same constant power, when the whole Universe is considered; and that the entropy grows while keeping the radius proportional to the inverse square of absolute temperature.

The research on the origin of Inertia is a problem that involved passionately a number of physicists especially from the crisis of Classical Physics and from the birth of General Relativity. Einstein himself underlined the importance of his meditation on this argument in the development of his theory of gravitation. Moreover he formulated precisely his reflections on inertia origin in the Mach's Principle - that in its simpler form says that inertia properties of matter are determined in some manner by the other bodies of the Universe. Even though with the development of General Relativity, Einstein rejected explicitly his first considerations on Inertia, the Principle represented, in a compact form, a research project that guided many scientists in the development of gravitational theories alternative to General Relativity (for example Brans-Dike Theory) - with the following cosmological implications - and the reformulations of Classical Mechanics based on Mach's reflections on Inertia (see for example the model proposed by Shrödinger).

ON BERRY'S MACHIAN ARGUMENT

Berry (1989) has posed a Machian query. Consider a body of mass m , acted on by a large one M located at a distance r , while the large mass has an acceleration $\vec{a}$ relative to the small one. In order to satisfy Mach's principle, the force exerted on the small mass by the larger, must contain a part proportional to $m\,\vec{a}$. By means of dimensional analysis, we find that the correct force should be proportional to $m\,\vec{a}$ and also to other terms: M, r, G and c, at some powers. According to Newton's third law, the power of M and m must be the same, i.e., they occur symmetrically in the force equation. The solution is,

$$\vec{F} = -GM\,\frac{m}{c^2 r}\,\vec{a}\,. \tag{8.5.2}$$

This looks like the force whose acceleration measures mutually accelerated charges. For the gravitational case, Sciama has given a name to it: law of inertial induction. For one thing, we may understand from the analogy, that if electromagnetic radiation is possible, then we would also have gravitational radiation.

If law (8.5.2) is to be applied to the distant masses of the Universe, in the Machian picture, and if Newton's second law should be valid, we need the following Brans-Dicke relation to be valid:

$$\frac{GM}{c^2 r} = 1\,, \tag{8.5.3}$$

so that,

$$\vec{F} = -GM\,\frac{m}{c^2 r}\,\vec{a} = -\,m\,\vec{a}\,. \tag{4}$$

This section was a digression on a Berry's argument. It must be said that from formula (8.5.4), we have the same kind of zero-total force applied to each and all particles in the Universe, (inertial force plus gravitational force, equals zero) so that we retrieve a zero-total energy of the Universe.

GRAVITATIONAL RADIATION IN SCIAMA'S MACHIAN MODEL

The Machian postulates are, *sphericity* (the Universe resembles a "ball" of approximate spherical shape), *egocentrism* (each observer sees the Universe from its center) and *democracy* (each point in space is equivalent to any other one – all observers are equivalent)(Berman, 2009).

Consider the rotating and expanding Universe. The angular speed is given by,

$$\omega = \pm\frac{c}{R}\,. \tag{8.5.5}$$

The reason for the above formula, is that, if we suppose that the Universe has constant zero-total energy, and is rotating, we would write the energy equation as:

$$E = 0 = Mc^2 - G\frac{M^2}{2R} + \frac{L^2}{MR^2}\,. \tag{8.5.6}$$

In the above, the inertial energy is represented by the first term in the r.h.s. of (8.5.6) followed by the potential energy and rotation terms, where L is the angular momentum of the Universe. We need such term in order to explain the Pioneer anomaly (Berman,

2007b), which was an anomalous constant deceleration which Berman says to be caused by the rotation of the Machian Universe, appearing as a centripetal one, and implying that the angular speed is given by relation (8.5.5) . The reason is that we find the Brans-Dicke "generalised" relations, which solve equation (8.5.6), namely,

$$G\frac{M}{c^2R} = \gamma \sim 1\ , \tag{8.5.1}$$

$$\frac{L^2}{c^2MR^2} = \gamma'\ , \tag{8.5.7}$$

where $\gamma' =$ constant, is that there is no other solution with time-invariant zero-energy equation (see equation 8.5.6), that allows time-varying $R = R(t)$. In fact, we find from (8.5.7) and (8.5.1), that $L \propto R^2$. On the other hand, we know from Newtonian physics that $L \cong RM\,(\omega R)$. From the two last relations, we find necessarily a relation of type (8.5.5) .

If we further calculate the power produced in the rotational motion of the "ball", we find,

$$P = \tau\omega = F_t R\omega = (M\alpha R)\,R\omega = MR^2\alpha\omega\ , \tag{8.5.8}$$

where τ is the torque of the tangential force F_t , which produces angular acceleration α .

From (8.5.5) we find,

$$\alpha = \frac{d\omega}{dt} = \frac{d}{dt}\left(\frac{c}{R}\right) = -cR^{-2}\dot{R} = -c^2R^{-2}, \tag{8.5.9}$$

where we have made use of the Machian relation for the radius of the causally connected Universe,

$$R = ct\ . \tag{8.5.10}$$

At last, we obtain,

$$P = -c^2M\omega = -\gamma\frac{c^5}{G}\ . \tag{8.5.11}$$

We would have obtained the same result, by introducing the equivalent of the electrodynamical Larmor power formula (Reitz et al, 1979), which yields the power radiated by an electric dipole, or an accelerated charge. The equivalent gravitational law is,

$$P_{Larmor} = \frac{2}{3}\left[\frac{GM^2}{c^3}\right]a^2\ , \tag{8.5.12}$$

where a is the acceleration. For the rotating case, $a = \omega^2 R$, and then, relation (8.5.12) becomes,

$$P_{Larmor} = \frac{2}{3}\left[\frac{GM^2}{c^3}\right]\omega^4R^2 = \frac{2\gamma^2}{3}\left[\frac{c^5}{G}\right]. \tag{8.5.13}$$

Both formulae ((8.5.11) and (8.5.13)) are pretty similar. It sounds as if Machian Universe is akin with Sciama's linearized gravitation. It is expected that the power-loss will be radiated as gravitational waves, as we shall discuss bellow.

QUADRUPOLE AND DIPOLE RADIATION

Einstein's quadrupole radiation formula (Weinberg, 1972), is of the type,

$$P_{einst} \approx \tfrac{G}{c^5}\omega^6 Q^2 \approx \tfrac{G}{c^5}\omega^6 \left[M^2 R^4\right] . \tag{8.5.14}$$

For our Machian Universe, the above will yield,

$$P_{einst} \approx \tfrac{c^5}{G}\gamma . \tag{8.5.15}$$

The formula for electrodynamical radiation depended on the dipole term's second time derivative, while the gravitational Einstein's radiation depends on the second time derivative of the quadrupole term. Nevertheless, for the Universe, we obtain the same result. Observe that we are in face of a constant radiating power, as was some time ago expressed by Berman (2008c). The fact, that for the Machian Universe, the dipole and the quadrupole power formulae coincide, does not mean that we may just take one instead of the other in local situations.

TEMPERATURE OF THE UNIVERSE. ENTROPY

The power of a black-body radiator is given by (Halliday, Resnick, Walker, 2008):

$$P_{bb} = \sigma A\, T^4 , \tag{8.5.16}$$

where A represents the radiating surface, at temperature T and σ is a constant.

For the Universe, we would find, on equating (8.5.16) with (8.5.11) ,

$$4\pi R^2 \sigma T^4 = \tfrac{c^5}{G} . \tag{8.5.17}$$

This results in the dependence of R with T^{-2} . This relation was found by Berman in several papers and books (see for instance, Berman, 2007, 2007a, 2008b). Now, let us calculate the entropy,

$$dS = \rho \left(4\pi R^2 dR\right) T^{-1} , \tag{8.5.18}$$

where ρ stands for the energy density of radiation, i.e.,

$$\rho = a\, T^4 , \tag{8.5.19}$$

so that,

$$S = \tfrac{4\pi}{3}\, a\, T^3\, R^3 \propto R^{\frac{3}{2}} . \tag{8.5.20}$$

We have found that the entropy of the Universe grows with $R^{\frac{3}{2}}$. This is a result of Sciama's theory, albeit Mach's theory. Berman has arrived to this formula in other cases (Berman, 2007, 2007a, 2008b, 2009).

CONCLUSIONS

Sciama's linear theory, has been expanded, through the analysis of radiating processes, thus, extending a previous paper of the present author (Berman, 2008d).

Larmor's power formula, in the gravitational version, leads to the correct constant power relation for the Machian Universe. However, we must remember that in local Physics, General Relativity deals with quadrupole radiation, while Larmor is a dipole formula; for the Machian Universe the resultant constant power is basically the same, either for our Machian analysis or for the Larmor and general relativistic formulae.

8.6. Exact Brans-Dicke Relation and Variable Speed of Light

The exact Brans-Dicke relation (Weinberg,1972), refers to the zero-total energy of the Universe, and states that,

$$\frac{GM}{c^2R} = \gamma \approx 1 \tag{8.6.1}$$

where R and M stand for the radius and mass of the causally related Universe, and $c(t)$ is the variable speed of light, while γ is a constant.

Berman (2007) proposed that given the fact that , from a theoretical point of view, the Universe had zero-total energy,when the gravitational interaction negative contribution was taken into account, one could estimate that the Universe should also have a spin.The numerical value of such spin, was then estimated, and pointed to an angular speed near ,

$$\omega \approx \pm\frac{c}{R} = 3.10^{-18}\text{rad.s}^{-1}. \tag{8.6.2}$$

This formula would match the experimentally measured retardation of Pioneers space-probes, launched by NASA more than thirty years ago, called the Pioneers Anomaly (Anderson et al.,2002).Later, Berman and Berman and Gomide(2011), retrieved this numerical relation in several theoretical frameworks.Several proofs of the zero-total energy were also produced, and the least modification that Robertson-Walker metric should suffer, in order to produce the rotating scenario, was then published (Berman,2008 a).Furthermore, it was shown that the same angular speed above, followed from the rotating metric, which kept all the achievements obtained up to now by the use of RW´s metric in Standard Cosmology (Berman and Gomide, 2011).This was an exact solution of Einstein´s equations.

If we take credit of the estimated angular speed above, (8.6.2), we obtain, $a \approx -\frac{c^2}{R} = -9.10^{-8}\text{cm.s}^{-2}$. (8.6.3)

This acceleration coincides with the Pioneers anomalous deceleration. The purpose of the present Section is to show how a variable speed of light, allows us to find a model of the Universe, coherent with Pioneers anomaly. A particular attempt by Shojaie(2010), leads to an approximate solution.We shall give an exact solution for the Pioneers Anomaly,with a more general model.

Rotation will not be the reason for the present solution, but it is evident that Berman and Gomide (2010;2011) solution employs rotation and General Relativity.

The usual Berman solution for constant deceleration paramenter is,

$$R = (mDt)^{1/m} \tag{8.6.4}$$

$$c = c_0 t^n \tag{8.6.5}$$

where m, n, c_0 and D are constants, and,

$$H = (mt)^{-1} \tag{8.6.6}$$

and the deceleration parameter q is given by,

$$q = m - 1 = -\frac{\ddot{R}R}{\dot{R}^2} \qquad (8.6.7).$$

The resultant mass solution is,

$$M = c_0^2 G^{-1} (mD)^{1/m} \; t^{2n+1/m} \tag{8.6.8}$$

We shall see bellow we build on the necessary deceleration.

PIONEERS ANOMALY SOLUTION

The observed Pioneers anomalous deceleration, coincides with,

$$\ddot{R} = -\frac{c^2}{R}$$

where the radius of the causally related Universe is given by 10^{28}cm.

We obtain exactly this relation, by specifying the conditions among constants of the above theory,

$$n + 1 = \frac{1}{m} \tag{8.6.9}$$

and,

$$c_0{}^2 = -\frac{n(mD)^{2/m}}{m^2(n+1)} \tag{8.6.10}$$

CONCLUSIONS

We found a general model for the Pioneers anomaly. On the other hand, in plain GRT, the solution of the above type is inexistent. Of course, the formalism we have employed is somehow equivalent to the rotational solution of Berman and Gomide (2010;2011). With the abundance of arbitrary constants, we may accommodate the experimental observations of present days. In particular, for accelerating models.

If we check for the spin of the Universe,and want to make it constant, by a variable speed of light, we suggest, while keeping the Pioneers deceleration,

$$n = -\frac{2}{3m} ,$$

so that,

$$L = MRc = \frac{c^3 R^2}{G} = \text{constant}$$

Otherwise,we could find that the mass obeys the constancy of the matter -energy, Mc^2 , by adjustment of constants,

$$1/m = -4n.$$

For the whole Universe, as expected,we find a cosmological-originated field of deceleration coincident with the Pioneers anomaly. As we still have room for other requirements,we may choose equations of state, deceleration or acceleration parameters, and we may even satisfy the constant matter energy requirement.

We praise prior research by Shojaie and Farhoudi(2004;2004a) and Shojaie(2010).

Chapter 9

Relativistic Cosmology and the Pioneers Anomaly

9.1. Generalised Robertson-Walker's Metric: Rotation Plus Expansion

Standard textbooks on Relativistic Cosmology, consider the usual Robertson-Walker's metric, which would represent a homogeneous and isotropic expanding Universe. However, the absence of rotation in the model, cast doubts on its validity as representing a real Universe: it would require a rigorous fine tuning, in order to keep, since inception, a non-rotating Universe. Since the Universe has been observed to be expanding with acceleration, the presence of a positive cosmological constant (Λ) , has been stressed in the last few years.

As a boundary condition, to be satisfied by Einstein's equations, the Machian condition has been put forward, sometimes in disguise. Berman (2007; 2007a; 2007b) has suggested that the consideration of a zero-total-energy Universe, might represent the Machian desired properties. Brans and Dicke (1961), presented new field equations that would satisfy the approximate relation,

$$\frac{GM}{c^2R} \sim 1 .$$

In the above, G represents the gravitational constant, and the mass M is that of a causally related Universe with radius R . Sabbata and Sivaram (1994), have shown that a closely related approximation would apply for the spin of the Universe L , namely,

$$\frac{L^2}{c^2M^2R^2} \sim 1.$$

We see that the Universe was expected to have a non-zero spin L . Though both relations above are heuristic, Berman has shown in the cited references, that the zero-total energy from the Newtonian point of view, could yield several exact relations which substitute the above two approximations. In conclusion: we expect a non-zero spin of the Machian Universe.

As we shall show below, by a simple expedient, we can obtain a rotational and expanding model, out of the original Robertson-Walker's metric.

Gomide and Uehara (1981) derived the field equations for a Robertson-Walker's metric in terms of coordinate time (t), when this time is not proper time (τ). In the most simplest case, we may write:

$$d\tau = (g_{00})^{1/2} dt \tag{9.1.1}$$

where,

$$g_{00} = g_{00}(t) . \tag{9.1.2}$$

The line element becomes:

$$ds^2 = -\frac{R^2(t)}{(1+kr^2/4)^2}\left[d\sigma^2\right] + g_{00}(t)\, dt^2. \tag{9.1.3}$$

The field equations, in General Relativity Theory (GRT) become:

$$3\dot{R}^2 = \kappa(\rho + \tfrac{\Lambda}{\kappa}) g_{00} R^2 - 3k g_{00}, \tag{9.1.4}$$

and,

$$6\ddot{R} = -g_{00}\kappa\left(\rho + 3p - 2\tfrac{\Lambda}{\kappa}\right) R - 3 g_{00} \dot{R}\, \dot{g}^{00} . \tag{9.1.5}$$

Local inertial processes are observed through proper time, so that the four-force is given by:

$$F^{\alpha} = \frac{d}{d\tau}(m u^{\alpha}) = m g^{00}\, \ddot{x}^{\alpha} - \frac{1}{2} m\, \dot{x}^{\alpha}\left[\frac{\dot{g}_{00}}{g_{00}^2}\right]. \tag{9.1.6}$$

Of course, when $g_{00} = 1$, the above equations reproduce conventional Robertson-Walker's field equations.

We must mention that the idea behind Robertson-Walker's metric is the Gaussian coordinate system. Though the condition $g_{00} = 1$ is usually adopted, we must remember that, the resulting time-coordinate is meant as representing proper time. If we want to use another coordinate time, we still keep the Gaussian coordinate properties.

From the energy-momentum conservation equation, in the case of a uniform Universe, we must have,

$$\frac{\partial}{\partial x^i}(\rho) = \frac{\partial}{\partial x^i}(p) = \frac{\partial}{\partial x^i}(g_{00}) = 0 \qquad (i = 1,2,3) . \tag{9.1.7}$$

The above is necessary in the determination of cosmic time, for a commoving observer. We can see that the hypothesis (9.1.2) – that g_{00} is only time-varying – is now validated.

In order to understand equation (9.1.6) , it is convenient to relate the rest-mass m , with an inertial mass M_i , with:

$$M_i = \frac{m}{g_{00}} \,. \tag{9.1.8}$$

It can be seen that M_i represents the inertia of a particle, when observed along cosmic time, i.e., coordinate time. In this case, we observe that we have two acceleration terms, which we call,

$$a_1^\alpha = \ddot{x}^\alpha, \tag{9.1.9}$$

and,

$$a_2^\alpha = -\frac{1}{2g_{00}} \left(\dot{x}^\alpha \dot{g}_{00}\right). \tag{9.1.10}$$

The first acceleration is linear; the second, resembles rotational motion.

If we consider a_2^α a centripetal acceleration, we conclude that the angular speed ω is given by,

$$\omega = \pm\frac{1}{2}\left(\frac{\dot{g}_{00}}{g_{00}}\right) . \tag{9.1.11}$$

We have both signs, because all that matters in the centripetal acceleration is the square of ω . However, there is a trick for the Universe.The fact that the Pioneers anomaly is a deceleration, does not mean that in the equation (9.1.5) above, we need a negative angular speed. This is a left-handed Universe.Because the equation can be written,

$$6\ddot{R}/R = -g_{00}\kappa\left(\rho + 3p - 2\tfrac{\Lambda}{\kappa}\right) + 6H\omega$$

and only a negative angular speed decelerates the last rhs term,we see that $\omega = -\frac{c}{R}$ gives the "right" deceleration term.But,if we consider the positive angular speed,$+c/R$,we can also find the correct deceleration of the Pioneers,through the formula $-\omega^2 R = -c^2/R$.What happens with the field equations is another story.General Relativity has two different solutions, here, unlike Newtonian theory.

If we now look at the prior chapter on Relativistic Cosmology, and take Raychaudhuri´s equation, one finds that the vorticity term of his equation, is repulsive always, so it is different than our negative angular speed case.

We may look to the above equation, as follows.

$$6\ddot{R} = 6\ddot{R}_\tau + 6\dot{R}\omega = (\text{standard RW´s result}) + 6\dot{R}\omega$$

We are NOT saying that the Pioneers anomalies depend on a negative $\ddot{R}$.In fact, large lambdas are expected in the present accelerating Universe, so that we shall find $\ddot{R} \geq 0$.

The conclusion is that we are not in face of a usual vorticity as in Raychaudhuri´s equation, but a new kind of rotation.The negative (or the positive) spin also explains the secondary Pioneers Anomaly, as was calculated in the last Chapter: the spacecraft is loosing spin, due to the negative contribution due to a cosmological origin. Each particle in the Universe shares a proportional part of the total spin of the Universe.

The first one has already been tackled by me.Now, let us face the second one.The spins of the Pioneers were telemetered, and as a surprise, it shows that the on-board measurements yield a decreasing angular speed, when the space-probes were not disturbed. Turyshev and Toth (2010), published the graphs (Figures 2.16 and 2.17 in their paper), from which it is clear that there is roughly an angular deceleration of about 0.1 RPM per three years, or,

$$\alpha \approx -1.2 \times 10^{-10} \mathrm{rad/s^2}.$$

As the diameter of the space-probes is about 10 meters, the linear acceleration is practically the Pioneers anomalous deceleration value ,in this case, -6.10^{-8} cm.s^{-2}.The present solution of the second anomaly, confirms our first anomaly explanation.

As the Universe expands, the cosmological component of the spin decreases. Of course, there may be other kinds of spin, but the cosmological effect is to decrease it.

By comparison between the usual τ – metric, and the field equations in the t – metric, we are led to conclude that the conventional energy density ρ and cosmic pressure p are transformed into $\bar{\rho}$ and $\bar{p}$, where:

$$\bar{\rho} = g_{00} \left(\rho + \frac{\bar{\Lambda}}{\kappa} \right) , \tag{9.1.12}$$

and,

$$\bar{p} = g_{00} \left(p - \frac{\bar{\Lambda}}{\kappa} \right) . \tag{9.1.13}$$

We plug back into the field equations, and find,

$$\bar{\Lambda} = \Lambda - \frac{3}{2\kappa} \left(\frac{\dot{R}}{R} \right) \dot{g}^{00} . \tag{9.1.14}$$

For a time-varying angular speed, considering an arc ϕ , so that,

$$\omega(t) = \frac{d\phi}{dt} = \dot{\phi} , \tag{9.1.15}$$

we find, from (9.1.11),

$$g_{00} = Ce^{2\phi(t)} \quad . \quad (C = \text{constant}) \tag{9.1.16}$$

Returning to (9.1.14), we find,

$$\bar{\Lambda} = \Lambda + \frac{3}{\kappa C} \left(\frac{\dot{R}}{R} \right) \omega e^{-2\phi(t)} . \tag{9.1.17}$$

We see that there is no singularity in the above relations for g_{00} and $\bar{\Lambda}$.

We have found an exact singularity-free solution, in GRT, for a rotating evolutionary Universe, derived from the original Robertson-Walker's metric, endowed with a lambda-term. The second right-hand-side term in (9.1.17) can be neglected as time passes by. The inertial mass M_i , depends on the rotation of the expanding Universe, fulfilling Mach's Principle.

The τ– field equations represent the compass of inertia; the t – metric, represents the rotation relative to the first one. By solving the field equations, one may find the cosmic pressure and energy density, for a given equation of state.

ZERO-TOTAL ENERGY-DENSITY OF THE ROTATING EVOLUTIONARY UNIVERSE

Consider the energy-density generalized Friedman-Robertson-Walker equation,

$$\rho - \frac{3H^2}{\kappa g_{00}} - \frac{3k}{\kappa R^2} + \frac{\Lambda}{\kappa} = 0$$

We write the above equation as,

$$\rho + \rho_{grav} + \rho_k + \rho_\Lambda \equiv \rho_{tot} = 0$$

where,

$\rho = \frac{Mc^2}{V}$ energy-density of matter.

$\rho_{grav} = -\frac{3H^2}{\kappa g_{00}}$ negative energy density of the gravitational field

$\rho_k = -\frac{3k}{\kappa R^2}$ energy density of the tricurvature

$\rho_\Lambda = \frac{\Lambda}{\kappa}$ energy density of the vacuum

$V \propto R^3$ volume

We have added all energy-densities that could possibly exist, and the sum is the total zero-valued energy density.as the rotational energy-density has not been includes in the sum, we need to imagine that it is included in the gravitational entry,because this is the only density that includes the $g_{00}(t)$ term.This is the reason why the total energy is also zero.We must take into consideration the negative energy and energy density of the gravitational field.As a consequence, there is no initial infinite energy density singularity. The total is zero.

The simplest proof of the zero-total energy of the rotating generalized R.W.s is left to Section 10.4 below.

Gödel's Universe (Adler et al., 1975), in which the matter does not uniquely determine the geometry, violates Mach's Principle, so that, there is not a distinguished universal time-coordinate in such Universe. More than that, the bulk matter of that Universe is commoving relative to a particular reference frame, but this frame is not inertial. The compass of inertia should rotate relative to the matter, or vice-versa, but the bulk matter represents the "fixed starts", so that, according to Mach's Principle, both can not rotate one relative to the other.

This is not the case in the present model. The background τ – metric defines the "fixed stars". The rotation becomes evident in the t – metric. All observers are commoving. This is done in a singularity-free framework.

The rotation of the Universe is a subject dealt by Berman (2007, 2007a, 2007b).

9.2. Generalised Robertson-Walker's Metric: Rotation Plus Expansion - II

Standard aspects concerning the treatment of rotation in General Relativity (GRT), were outlined by Islam (1985). Rotation of the Universe was dealt by Berman from several angles, like the Machian (Berman, 2007; 2007a; 2007b; 2007c) and under the Robertson-Walker's metric (Berman, 2008; 2008a). We now shall be dealing with the cosmological metric that would be fitted for the study of a rotating and expanding Universe, in a parallel approach with new kinds of arguments.

It is usual and practical, to employ metrics "adapted" to solve a given physical problem. Any metric induces a topology. The same topology, nevertheless, can be met by other less adapted metrics. The whole set of metrics, can be used to solve any particular problem. According to Synge (1960), what matters is only curvature in each given point, and not, for instance, a typical observer's proper quadri-acceleration along his world-line, for some metric. Joshi (1993) comments that GRT restricts spacetime, according to the principle of local flatness and its Special Relativistic framework. However, local restrictive physics, do not affect global topology of the spacetime manifold. Boundary conditions, to the contrary, limit the possible topologies. One such condition is revealed by Mach's Principle, and may satisfy macroscopical approaches, but for microphysics, a more elaborated topology must be found. In a private communication, A. Sant'Anna, has pointed out that micro-topologies are now under research, in order to produce viable scenarios in Quantum computer theory.

Without delving in the above related details, I showed elsewhere, that for a GRT rotating model, Robertson-Walker's metric could be generalised in order to fit the problem of rotation plus expansion. This will be the subject of our present Section.

ROTATING EVOLUTIONARY METRICS

When we are not working with proper time τ, but make a transformation to any other time-coordinate t. Generally, instead, we write:

$$d\tau^2 \equiv g_{00}(r,\theta,\phi,t)\, dt^2 \, . \tag{9.2.1}$$

The Robertson-Walker's metric is written usually in terms of proper time, namely,

$$ds^2 = d\tau^2 - \frac{R^2}{\left(1+k\frac{r^2}{4}\right)^2} d\sigma^2 \,. \tag{9.2.2}$$

If we change from the angular coordinate ϕ towards $\tilde{\phi}$, such that:

$$\tilde{\phi} = \phi - \omega\, t \,, \tag{9.2.3}$$

or,

$$d\tilde{\phi} = d\phi - \omega\, dt \,, \tag{9.2.3a}$$

we shall find that tri-dimensional metric element becomes,

$$\begin{aligned} d\,\tilde{\sigma}^2 &= dr^2 + r^2 d\theta^2 + r^2 \sin^2\theta\,(d\phi - \omega dt)^2 = \\ &= d\sigma^2 + r^2 \sin^2\theta\;\omega^2 dt^2 - 2\omega\, r^2 \sin^2\theta\; d\phi\; dt \,. \end{aligned} \tag{9.2.4}$$

The new metric is of the following form:

$$d\,\tilde{s}^2 = g_{00} dt^2 - \frac{R^2}{\left(1+k\frac{r^2}{4}\right)^2}\left[d\sigma^2 - 2\omega r^2 \sin^2\theta d\phi dt\,\right], \tag{9.2.5}$$

with,

$$g_{00} = 1 - \frac{\omega^2 R^2 r^2 \sin^2\theta}{\left(1+k\frac{r^2}{4}\right)^2} \,. \tag{9.2.5a}$$

Consider now a totally comoving observer: his proper time is given by,

$$d\tau^2 = g_{00} dt^2 = \left[1 - \varpi^2 R^2\right] dt^2, \tag{9.2.6}$$

where

$$\varpi = \pm \frac{\omega r \sin\theta}{\left(1+k\frac{r^2}{4}\right)} \,. \tag{9.2.7}$$

If we want to preserve $g_{00} > 0$, so that equation (9.2.6) represents real proper time, we may solve the problem with,

$$\varpi = \pm \frac{\alpha}{R} \,, \tag{9.2.8}$$

with $\alpha^2 < 1$.

For instance, we could fix $\dot{\alpha} = 0$ so that, $\alpha = \alpha(r)$ then we would also find,

$$\omega = \pm \frac{\beta}{R} \,,$$

with a similar condition, $\dot{\beta} = 0$, and $\beta = \beta(r)$.

We call ϖ as the effective angular speed of the Universe; it has a striking similarity with Berman's solution for a Machian rotating Universe (Berman, 2007b). We notice that such angular speed causes a kind of centripetal ubiquitous acceleration having a universal character and which caused the so-called Pioneer anomaly, in two space probes, launched by NASA more than thirty years ago. By the same token, we define a tangential speed ,

$$V = \varpi R < 1 \ .$$

In order to make contact with usual cosmological theory, we consider now that, the non-diagonal metric term which points towards a Universal precession of gyroscopes, is in the same token, as in the Lense-Thirring metric. Consider now a semi-comoving observer, defined by the condition $d\tilde{\phi} = 0$. If we write,

$$\omega = \frac{d\phi}{dt} \ ,$$

the non-diagonal term becomes diagonalized, like,

$$2\frac{\varpi^2}{\omega}R^2 d\phi \, dt = 2\varpi^2 R^2 dt^2 \ . \tag{9.2.9}$$

In this case, the apparent temporal metric coefficient becomes,

$$(g_{00})_{ap} = 1 - \varpi^2 R^2 + 2\varpi^2 R^2 = 1 + \varpi^2 R^2 \ . \tag{9.2.10}$$

With the value of ϖ given by relation (9.2.8) the apparent temporal metric coefficient becomes time-independent. Of course, the real temporal coefficient is then also so.

GENERALISED GAUSSIAN METRICS

We refer to Berman (2007; 2007a), for a presentation of the Gaussian metrics which does not suffer from the "trap" of considering $g_{00} =$ constant for their definition. What really matters in a Gaussian metric, is that the time axis is orthogonal to the tri-space. When such metric represents a tri-space which is not into rotation around the time axis, coordinate time means proper time τ . Now relax the restriction, and consider that the orthogonal tri-space is rotating around the time axis: the time coordinate that we must use is t , defined by (9.2.1) . This was indeed shown in the previous section, and points to what we shall call Generalised Gaussian metrics.

Berman's definition of Gaussian metrics (Berman 2007; 2007a), constrains only the metric by the condition,

$$g^{ij}\frac{\partial g_{i0}}{\partial t} = 0 \qquad (i, j = 1, 2, 3) \ . \tag{9.2.11}$$

On the other hand, Gaussian normal coordinates, are defined by the condition,

$$g_{i0}(t) = 0 \ . \tag{9.2.12}$$

Then, the following condition applies for a commoving observer:

$$g_{00}u^0 = 1 , \tag{9.2.13}$$

where, u^0 is the temporal component of the quadri-velocity.

From the cosmological point of view, it has been suggested that rotation of the Universe is associated with cosmic microwave background radiation's quadrupole anisotropies. These have not been significant: this may be attributed to low angular speeds, less than what is possibly measured by present technology. We must remember that the cosmic no-hair conjecture is really established by means of an inflationary phase erasing the angular speed (remember equation (9.2.8), with R exponentially increasing).

Another point is that CMBR analyses, only apply to the equation of null geodesics, $ds = 0$. To the contrary, the Pioneer anomaly deals with $ds \neq 0$. It must be stressed that a variable temporal metric coefficient has been studied long ago by Gomide and Uehara (1981).

SOME COMMENTS

The rotation of the Universe, not only explains the Pioneers' anomaly, but would be in the right direction, in order to explain the left handed preference of neutrinos' spins, parity violations and the related matter-antimatter asymmetry (Feynman et al., 1965). It has been

found that the DNA helix is left handed. Our bodies are not symmetric; molluscs have likewise shells; aminoacids in living bodies, too (Barrow and Silk, 1983).

We predict that chaotic phenomena, and fractals, in the Universe, as well as rotations of Galaxies and their clusters, must have a predilection towards the left hand. We predict that the directions of the magnetic field of the Universe, and rotation, are related with the laws of electrodynamics and the left-hand.

It is important to note the preliminary results on the Universe rotation, by Birch (1982; 1983) and Gomide, Berman and Garcia (1986). Sciama's inertial theory (Sciama, 1953), also contemplated a rotation speed of the type given by us (see (9.2.8)). The spin of the Universe is a subject of two recent, and a forthcoming, papers by Berman (2007b; 2007c; 2008). Machian rotations are dealt by Berman (2007; 2007a).

I predict that with improving technological tools, the rotation of the Universe will be experimentally measured in the future.

9.3. The Pioneers Anomaly and a Machian General Relativistic Model

Ni (2008;2009), has reported observations on a possible rotation of the polarization of the cosmic background radiation, around 0.1 radians.As such radiation was originated at the inception of the Universe, we tried to estimate a possible angular speed or vorticity, by dividing 0.1 radians by the age of the Universe , obtaining about 10^{-19}rad.s^{-1}.

The numerical result is very close to the theoretical estimate, by Berman (2007),

$$\omega \approx \pm c/R = 3.10^{-18}\text{rad.s}^{-1}.$$

where c , R represent the speed of light in vacuum, and the radius of the causally related Universe.

We must remember, as Berman and Gomide (2011) have pointed, that their calculation deals with material particles, or, in the language of General Relativity, non-null geodesics.The fact that the Universe may exhibit a rotating state, can be understood by a simple fine-tuning argument—it would be highly improbable that the Universe could keep since birth a state of no angular momentum at all.

The value of Berman´s rotation, fits with the Pioneers anomaly, which consists on decelerations suffered by NASA space probes in non- closed curves, extending to outer space.

About this same numerical value of the angular speed is predicted also in Gödel´s rotational model, but it is not an expanding one(see Adler, Bazin and Schiffer,1975). In the next few years, the observational evidence may confirm or not such rotation .

Rotating metrics in General Relativity were first studied by Islam (1985), but Cosmology was not touched upon. However, it would be necessary an extreme perfect fine-tuning, in order to create the Universe without any angular-momentum. The primordial Quantum Universe, is characterized by dimensional combinations of the fundamental constants "c" , "h" and "G" respectively the speed of light in vacuo, Planck's and Newton's gravitational constants. The natural angular momentum of Planck's Universe, as it is called, is, then, "h" . It will be shown that the angular momentum grows with the expanding Universe, but the corresponding angular speed decreases with the scale-factor (or radius) of the Universe, such being the reason for the difficulty in detection of this speed with present technology. Notwithstanding, the so-called Pioneers' anomaly (Anderson,2002), which is a deceleration verified in the Pioneers space-probes launched by NASA more than thirty years ago, was attributed by Berman, to a "Machian" ubiquitous field of centripetal accelerations, due to the rotation of the Universe. Berman's calculation rested on the assumption that the zero-total energy of the Universe was a valid result for the rotating case, but the proof was not supplied in that paper (Berman, 2007b). By "proof", one thinks on the pseudotensor energy calculations of General Relativity — the best gravitational theory ever published.

In his three best-sellers (Hawking, 1996; 2001; 2003), Hawking describes inflation (Guth, 1981; 1998), as an accelerated expansion of the Universe, immediately after the creation instant,while the Universe, as it expands,borrows energy from the gravitational field to create more matter. According to his description, the positive matter energy is exactly balanced by the negative gravitational energy, so that the total energy is zero,and that when the size of the Universe doubles, both the matter and gravitational energies also double, keeping the total energy zero (twice zero).Moreover, in the recent, next best-seller,Hawking and Mlodinow(2010) comment that if it were not for the gravity interaction, one could not validate a zero-energy Universe, and then, creation out of nothing would not have happened. On the other hand, Berman (2008a; b) has shown that Robertson-Walker's metric, is a particular, non-rotating case, of a general relativistic expanding and rotating metric first developed by Gomide and Uehara (1981). The peculiarity of the general metric is that instead of working with proper-time τ , one writes the field equations of General Relativity with a cosmic time t related by:

$$d\tau = (g_{00})^{1/2} dt \,, \tag{9.3.1}$$

where,

$$g_{00} = g_{00}(r, \theta, \phi, t) \,. \tag{9.3.2}$$

It will be seen that when one introduces a metric temporal coefficient g_{00} which is not constant, the new metric includes rotational effects. In a previous paper Berman (2009c) has calculated the energy of the Friedman-Robertson-Walker's Universe, by means of pseudo-tensors, and found a zero-total energy. **Our main task will be to show why the Universe is a zero-total-energy entity, by means of pseudo-tensors, even when one chooses a variable g_{00} such that the Universe also rotates, and then, to show how General Relativity predicts a universal angular speed, and a universal centripetal deceleration, numerically coincident with the observed deceleration of the Pioneers space-probes.** The first calculation of this kind, with the Gomide -Uehara generalization of RW´s metric, was undertaken by Berman (1981), in his M.Sc. thesis, advised by the present second author, but where the rotation of the Universe was not the scope of the thesis.

We shall argue below that, for the Universe, local and global Physics blend together. In a previous paper (Berman, 2009c), we stated that "if the Universe has some kind of rotation, the energy-momentum calculation refers to a co-rotating observer". Such being the case, we now go ahead for the actual calculations, involving rotation. Birch (1982; 1983) cited inconclusive experimental data on a possible rotation of the Universe, which was followed by a paper written by Gomide, Berman and Garcia (1986).The field equations were presented above.

The angular speed, is given by,

$$\omega = \pm \tfrac{1}{2} \left(\frac{\dot{g}_{00}}{g_{00}} \right) . \tag{9.1.11}$$

The case where g_{00} depends also on r, θ and ϕ was considered also by Berman (2008b) and does not differ qualitatively from the present analysis, so that, we refer the reader to that paper.

ENERGY OF THE ROTATING EVOLUTIONARY UNIVERSE

Even in popular Science accounts(Hawking,1996; 2001; 2003;— and Moldinow,2010;

Guth,1998), it has been generally accepted that the Universe has zero-total energy. The first such claim, seems to be due to Feynman(1962-3). Lately, Berman(2006, 2006 a) has proved this result by means of simple arguments involving Robertson-Walker's metric for any value of the tri-curvature ($0, -1, 1$).A similar simple proof will be given in Section10.4, for the rotating metric

The pseudotensor t_ν^μ , also called Einstein's pseudotensor, is such that, when summed with the energy-tensor of matter T_ν^μ , gives the following conservation law:

$$\left[\sqrt{-g}\,(T_\nu^\mu + t_\nu^\mu)\right]_{,\mu} = 0\,. \tag{9.3.3}$$

In such case, the quantity

$$P_\mu = \int \left\{\sqrt{-g}\left[T_\mu^0 + t_\mu^0\right]\right\} d^3x\,, \tag{9.3.4}$$

is called the general-relativistic generalization of the energy-momentum four-vector of special relativity (Adler et al, 1975).

It can be proved that P_μ is conserved when:

a) $T_\nu^\mu \neq 0$ only in a finite part of space; and,
b) $g_{\mu\nu} \longrightarrow \eta_{\mu\nu}$ when we approach infinity, where $\eta_{\mu\nu}$ is the Minkowski metric tensor.

However, there is no reason to doubt that, even if the above conditions were not fulfilled, we might eventually get a constant P_μ , because the above conditions are sufficient, but not strictly necessary. We hint on the plausibility of other conditions, instead of a) and b) above.

Such a case will occur, for instance, when we have the integral in (9.3.4) is equal to zero.

For our generalised metric, we get exactly this result, because, from Freud's (1939) formulae, there exists a super-potential, (Papapetrou, 1974):

$${}_F U_\lambda^{\mu\nu} = \frac{g_{\lambda\alpha}}{2\sqrt{-g}}\left(\bar{g}^{\mu\alpha}\bar{g}^{\nu\beta} - \bar{g}^{\nu\alpha}\bar{g}^{\mu\beta}\right)_{,\beta}\,,$$

where the bars over the metric coefficients imply that they are multiplied by $\sqrt{-g}$, and such that,

$$\kappa\sqrt{-g}(T_\lambda^\rho + t_\lambda^\rho) = {}_F U_\lambda^{\rho\sigma}{}_{,\sigma},$$

thus finding, after a brief calculation, for the rotating Robertson-Walker's metric,

$$P_\lambda = 0\,.$$

The above result, with von Freud's superpotential, which yields Einstein's pseudotensorial results, points to a zero-total energy Universe, even when the metric is endowed with a varying metric temporal coefficient .

A similar result would be obtained from Landau-Lifshitz pseudotensor (Papapetrou, 1974), where we have:

$$P^{\nu}_{LL} = \int (-g) \left[T^{\nu 0} + t^{\nu 0}_{L} \right] \, d^3x, \tag{9.3.5}$$

where,

$$\kappa\sqrt{-g}(T^{\mu\rho} + \tilde{t}^{\mu\rho}) = \tilde{U}^{\mu\rho\sigma}{}_{,\sigma} \, ,$$

and,

$$\tilde{U}^{\mu\rho\sigma} = \bar{g}^{\lambda\mu} \, {}_{F}U^{\rho\sigma}_{\lambda} \, ,$$

A short calculation shows that, for the rotating metric,too, we keep valid the result,

$$P^{\nu}_{LL} = 0 \qquad (\nu = 0, 1, 2, 3 \ ) \quad . \tag{9.3.6}$$

Other superpotentials would also yield the same zero results. A useful source for the main superpotentials in the market, is the paper by Aguirregabiria et al. (1996).

The equivalence principle, says that at any location, spacetime is (locally) flat, and a geodesic coordinate system may be constructed, *where the Christoffel symbols are null. The pseudotensors are, then, at each point, null. But now remember that our old Cosmology requires a co-moving observer at each point. It is this co-motion that is associated with the geodesic system, and, as RW´s metric is homogeneous and isotropic, for the co-moving observer, the zero-total energy density result, is repeated from point to point, all over spacetime. Cartesian coordinates are needed, too, because curvilinear coordinates are associated with fictitious or inertial forces, which would introduce inexistent accelerations that can be mistaken additional gravitational fields (i.e.,that add to the real energy). Choosing Cartesian coordinates is not analogous to the use of center of mass frame in Newtonian theory, but the null results for the spatial components of the pseudo-quadrimomentum show compatibility.*

AN ALTERNATIVE DERIVATION

Though so many researchers have dealt with the energy of the Universe, our present original solution involves rotation. We may paraphrase a previous calculation, provided that we work with proper time τ instead of coordinate time t (Berman, 2009c). Then, the rotation of the Universe will be automatically included. We shall now consider, first, why the Minkowski metric represents a null energy Universe . Of course, it is empty. But, why it has zero-valued energy? We resort to the result of Schwarzschild´s metric, (Adler et al., 1975), whose total energy is,

$$E = Mc^2 - \frac{GM^2}{2R}$$

If $M = 0$, the energy is zero,too. But when we write Schwarzschild´s metric, and make the mass become zero, we obtain Minkowski metric, so that we got the zero-energy result. Any flat RW´s metric, can be reparametrized as Minkowski´s; or,for closed and open Universes, a superposition of such cases (Cooperstock and Faraoni,2003; Berman, 2006; 2006a).

Now, the energy of the Universe, can be calculated at constant time coordinate τ . In particular, the result would be the same as when $\tau \to \infty$, or, even when $\tau \to 0$. Arguments

for initial null energy come from Tryon(1973), and Albrow (1973). More recently, we recall the quantum fluctuations of Alan Guth´s inflationary scenario (Guth,1981;1998). Berman (see for instance,2008c), gave the Machian picture of the Universe, as being that of a zero energy . Sciama´s inertia theory results also in a zero-total energy Universe (Sciama, 1953; Berman, 2008d;2009e).

Consider the possible solution for the rotating case. We work with the τ-metric, so that we keep formally the RW´s metric in an accelerating Universe. The scale-factor assumes a power-law , as in constant deceleration parameter models (Berman,1983;—and Gomide,1988),

$$R = (mD\tau)^{1/m} , \tag{9.3.7}$$

where, m , $D =$ constants, and,

$$m = q + 1 > 0 \tag{9.3.8}$$

where q is the deceleration parameter.

For a perfect fluid energy tensor, and a perfect gas equation of state, cosmic pressure and energy density obey the following energy-momentum conservation law, (Berman, 2007,

2007a),

$$\dot{\rho} = -3H(\rho + p) , \tag{9.3.9}$$

where, only in this Section, overdots stand for τ-derivatives .Let us have,

$$p = \alpha\rho \qquad (\alpha = \text{ constant larger than } -1) . \tag{9.3.10}$$

On solving the differential equation, we find, for any $k = 0$, 1 , -1 ,that,

$$\rho = \rho_0 \tau^{-\frac{3(1+\alpha)}{m}} \qquad (\rho_0 = \text{ constant}) . \tag{9.3.11}$$

When $\tau \to \infty$, from (9.3.11) we see that the energy density becomes zero, and we retrieve an "empty" Universe, or, say, again, the energy is zero. However, this energy density is for the matter portion, but nevertheless, as in this case, $R \to \infty$, all masses are infinitely far from each others, so that the gravitational inverse-square interaction is also null. The total energy density is null, and, so, the total energy. Notice that the energy-momentum conservation equation does not change even if we add a cosmological constant density, because we may subtract an equivalent amount in pressure, and equation (9.3.9) remains the same. The constancy of the energy, leads us to consider the zero result at infinite time, also valid at any other instant.

We refer to Berman (2006; 2006a) for another alternative proof of the zero-energy Universe. If we took τ instead of t, these references would provide the zero result also for the rotational case.

PIONEERS ANOMALY REVISITED

Einstein's field equations (9.1.4) and (9.1.5) above, can be obtained, when $g_{00} =$ constant, through the mere assumptions of conservation of energy (equation 9.1.4) and thermodynamical balance of energy (equation 9.1.5), as was pointed out by Barrow (1988). The latter is also to be regarded as a definition of cosmic pressure, as the volume derivative of energy with negative sign ($p = -\frac{d(\rho V)}{dV}$).

Now, let us consider a time-varying g_{00}. We may write the energy (in fact, the "energy-density")– equation, as follows:

$$\frac{3\dot{R}^2}{g_{00}} - \kappa(\rho + \frac{\Lambda}{\kappa})R^2 = -3k = \text{ constant} . \tag{9.3.12}$$

The r.h.s. stands for a constant. We can regard the l.h.s. as the a sum of constant terms, thus finding a possible solution of the field equations, such that each term in the l.h.s. of (9.3.12) remains constant. For example, let us consider, the Machian kind of solution below,

$$\rho = \rho_0 R^{-2} , \tag{9.3.13}$$

$$\Lambda = \Lambda_0 R^{-2}, \tag{9.3.14}$$

$$g_{00} = 3\gamma^{-1}\dot{R}^2, \tag{9.3.15}$$

where, ρ_0 , Λ_0 and γ are non-zero constants.

When we plug the above solution to the cosmic pressure equation (9.1.5), we find that it is automatically satisfied provided that the following conditions hold,

$$2\Lambda_0 = \kappa\rho_0(1+3\alpha) , \tag{9.3.16}$$

$$p = \alpha\rho \qquad (\alpha = \text{constant}) , \tag{9.3.17}$$

and,

$$\gamma = \kappa\rho_0 + \Lambda_0 - 3k \tag{9.3.17a}$$

As we found a general-relativistic solution, so far, we are entitled to the our previous general relativistic angular speed formula (9.1.11), to which we plug our solution (9.3.15), to wit,

$$\pm\omega = \frac{\ddot{R}}{\dot{R}} = H + \frac{\dot{H}}{H}.$$

For the power-law solution of the last Section,

$$H = \frac{1}{mt} ,$$

so that,if we choose the negative sign for the angular speed,

$$-\omega = -\frac{q}{mt} \approx t^{-1},$$

where we roughly estimated the present deceleration parameter as $-1/2$, while, the age is estimated as 10 billion years.

The centripetal acceleration,

$$a = -\omega^2 R \approx -t^{-2} R \simeq -9.10^{-8} \text{ cm.s}^{-2}.$$

Notice that the same result would follow from a scale-factor varying linearly with time. This is the sort of scale-factor associated with the Machian Universe. In fact,the field equations that we had (equations (9.1.4) and (9.1.5)), were not enough in order to determine the exact form of the scale-factor, because we had an extra-unknown term, the temporal metric coefficient. When we advance a given equation of state, the original RW´s field equations, with constant g_{00},may determine the scale-factor´s formula. Just to remember, our solution is a particular one.

This is a general relativistic result. It matches Pioneers anomalous deceleration.We need a negative angular speed, in order that the contribution of the angular speed be negative, for $\ddot{R}$.

If we would like to consider an age of, say, 13 billion years,we would need to take a deceleration parameter $q = -6/11$.

If we calculate the centripetal acceleration corresponding to the above angular speed, we find, for the present Universe, with $R \approx 10^{28}$cm and $c \simeq 3.10^{10}$cm.s^{-2} ,

$$a_{cp} = -\omega^2 R \cong -9.10^{-8} cm/s^2$$

This value matches the observed experimentally deceleration of the NASA Pioneers' space-probes.

We observe that the Machian picture above is understood to be valid for any observer in the Universe, i.e., the center of the "ball" coincides with any observer; the "Machian" centripetal acceleration should be felt by any observed point in the Universe subject to observation from any other location.

We solve also other mystery concerning Pioneers anomaly. It has been verified experimentally, that those space-probes in closed (elliptical) orbits do not decelerate anomalously, but only those in hyperbolic flight. The solution of this other enigma is easy, according to our view. The elliptical orbiting trajectories are restricted to our local neighborhood, and do not acquire cosmological features, which are necessary to qualify for our Machian analysis, which centers on cosmological ground. But hyperbolic motion is not bound by the Solar system, and in fact those orbits extend to infinity, thus qualifying themselves to suffer the cosmological Machian deceleration.

SOME COMMENTS AND DISCUSSION

Someone has made very important criticisms on our work.First, he says why do not the planets in the solar system show the calculated deceleration on the Pioneers? The reason is that elliptical orbits are closed, and localized. You do not feel the expansion of the universe in the sizes of the orbits either. In General Relativity books, authors make this explicit.You do not include Hubble´s expansion in Schwarzschild´s metric.But, those space probes that undergo hyperbolic motion, which orbits extend towards infinity, they acquire cosmological characteristics, like, the given P.A. deceleration.Second objection, there are important papers which resolve the P.A. with non-gravitational Physics. The answer,— that is OK, we have now alternative explanations.This does not preclude ours.Third, cosmological reasons were discarded, including rotation of the Universe. The problem is that those discarded cosmologies, did not employ the correct metric.For instance, they discarded rotation by examining Gödel model, which is non expanding, and with a strange metric. The kind of metric we employ now, or the one that we employed in the rotational case, were not discarded or discussed by the authors cited by this objecter. Then, the final question, is how come that the well respected author L.Iorio (2010) dismissed planetary Coriolis forces induced by rotation of distant masses, by means of the constraints in the solar system. Our answer is that, beside what we answered above, he needs to consider Mach´s Principle on one side, and the theoretical meaning of vorticities, because one is not speaking in a center or an axis of rotation or so.When we say, in Cosmology, that the Universe rotates, we mean that there is a field of vorticities,just that.The whole idea is that Cosmology does not enter the Solar System except for non-closed orbits that extend to outer space.We ask the reader to check Mach´s Principle, because in some formulations of this principle, rotation is in fact a *forbidden affaire*.

Another one pointed out a different "problem". He objects, that the angular speed formula of ours, is coordinate dependent.Now, when you choose a specific metric, you do it thinking about the kind of problem you have to tackle. After you choose the convenient metric, you forget tensor calculus, and you work with coordinate-dependent relations.They work only for the given metric, of course.

We have obtained a zero-total energy proof for a rotating expanding Universe. The zero result for the spatial components of the energy-momentum-pseudotensor calculation, are equivalent to the choice of a center of Mass reference system in Newtonian theory, likewise the use of co-moving observers in Cosmology. It is with this idea in mind, that we are led to the energy calculation, yielding zero total energy, for the Universe, as an acceptable result: we are assured that we chose the correct reference system; this is a response to the criticism made by some scientists which argue that pseudotensor calculations depend on the reference system, and thus, those calculations are devoid of physical meaning.

Related conclusions by Berman should be consulted (see all Berman's references at the end of this article). As a bonus, we can assure that there was not an initial infinite energy density singularity, because attached to the zero-total energy conjecture, there is a zero-total energy-density result, as was pointed by Berman elsewhere (Berman, 2008).The so-called total energy density of the Universe, which appears in some textbooks, corresponds only to the non-gravitational portion, and the zero-total energy density results when we subtract from the former, the opposite potential energy density.

As Berman(2009d; f) shows, we may say that the Universe is *singularity -free*, and was created *ab-nihilo*, nor there is zero-time infinite energy-density singularity.

Paraphrasing Dicke (1964; 1964a), it has been shown the many faces of Dirac's LNH, as many as there are about Mach's Principle. In face of modern Cosmology, the naif theory of Dirac is a foil for theoretical discussion on the foundations of this branch of Physical theory. The angular speed found by us,(Berman,2010;2009a), matches results by Gödel (see Adler et al., 1975), Sabbata and Gasperini (1979), and Berman (2007b, 2008b, c).

Rotation of the Universe and zero-total energy were verified for Sciama's linear theory, which has been expanded, through the analysis of radiating processes, by one of the present authors (Berman, 2008d;2009e).There,we found Larmor's power formula, in the gravitational version, leads to the correct constant power relation for the Machian Universe. However, we must remember that in local Physics, General Relativity deals with quadrupole radiation, while Larmor is a dipole formula; for the Machian Universe the resultant constant power is basically the same, either for our Machian analysis or for the Larmor and general relativistic formulae.

Referring to rotation, it could be argued that cosmic microwave background radiation deals with null geodesics, while Pioneers' anomaly, for instance, deals with time-like geodesics. In favor of evidence on rotation, we remark neutrinos' spin, parity violations, the asymmetry between matter and anti-matter, left-handed DNA-helices, the fact that humans and animals alike have not symmetric bodies, the same happening to molluscs. And, of course, the results of the rotation of the polarization of CMBR.

We predict that chaotic phenomena and fractals, rotations in galaxies and clusters, may provide clues on possible left handed preference through the Universe.

Berman and Trevisan (2010) have remarked that creation out-of-nothing seems to be supported by the zero-total energy calculations. Rotation was now included in the derivation of the zero result. We could think that the Universes are created in pairs, the first one (ours), has negative spin and positive matter; the second member of the pair, would have negative matter and positive spin: for the ensemble of the two Universes, the total mass would always be zero; the total spin, too. The total energy (twice zeros) is also zero.

9.4. General Relativistic Cosmological Models with Pioneers Anomaly

The Pioneers Anomaly is the deceleration of about -9.10^{-8}cm.s^{-2} suffered by NASA space-probes travelling towards outer space.It has no acceptable explanation within local Physics, but if we resort to Cosmology, it could be explained by the rotation of the Universe.Be cautious, because there is no center or axis of rotation.We are speaking either of a Machian or a General Relativistic cosmological vorticity.It could apply to each observed point in the Universe, observed by any observer.Another explanation, would be that our Universe obeys a variable speed of light Relativistic Cosmology, without vorticities.However, we shall see later that both models are equivalent.

In previous papers (Berman and Gomide,2010;2011), by considering an exact but particular solution of Einstein´s field equations for an expanding and rotating metric,found, by estimating the deceleration parameter of the present Universe, as $q \approx -1/2$, that the Universe appeared to possess a field of decelerations coinciding approximately with the Pioneers anomalous value (Anderson et al., 2002).We now shall consider the condition for an exact match with the Pioneers deceleration,with a large class of solutions in General Relativity.

Ni (2008;2009), has reported observations on a possible rotation of the polarization of the cosmic background radiation, around 0.1 radians.As such radiation was originated at the inception of the Universe, we tried to estimate a possible angular speed or vorticity, by dividing 0.1 radians by the age of the Universe , obtaining about 10^{-19}rad.s^{-1}.

The numerical result is very close to the theoretical estimate, by Berman (2007),

$$\omega = \pm c/R = 3.10^{-18}\text{rad.s}^{-1}. \tag{9.4.1}$$

where c,R represent the speed of light in vacuum, and the radius of the causally related Universe.

When one introduces a metric temporal coefficient g_{00} which is not constant, the new metric includes rotational effects. The metric has a rotation of the tri-space (identical with RW´s tri-space) around the orthogonal time axis.The field equations were already given.

$$\omega = \pm\frac{1}{2}\left(\frac{\dot{g}_{00}}{g_{00}}\right). \tag{9.4.2}$$

The case where g_{00} depends also on r, θ and ϕ was considered also by Berman (2008b) and does not differ qualitatively from the present analysis, so that, we refer the reader to that paper.

AN EXACT SOLUTION TO THE PIONEERS ANOMALY

Consider the possible solution for the rotating case.We equate (9.4.1) and (9.4.2).We try a power-law solution for R,and find,

$g_{00} = Ae^{\pm t^{1-1/m}}$ (A =constant).The positive exponential is for positive angular speed, and vice-versa.

The scale-factor assumes a power-law , as in constant deceleration parameter models (Berman,1983;—and Gomide,1988),

$$R = (mDt)^{1/m}\,, \tag{9.4.3}$$

where, m , $D =$ constants, and,

$$m = q + 1 > 0, \tag{9.4.4}$$

where q is the deceleration parameter. We may choose q as needed to fit the observational data.

We find,

$$H = (mt)^{-1}$$

If we now solve for energy-density of matter, and cosmic pressure, for a perfect fluid,the best way to present the calculation, and the most simple, is showing the matter energy-densityρ and the σ-or gravitational density parameter, to be defined below. We find, $\rho = \frac{3t^{-2}}{m^2 A\kappa} e^{\pm t^{1-1/m}} + 3k(mDt)^{-2/m} - \frac{\Lambda}{\kappa}$ where the negative exponential corresponds to positive angular speed and vice-versa.

$$\sigma = \left(\rho + 3p - 2\tfrac{\Lambda}{\kappa}\right) = 6\left[\frac{(1-m)t^{-2} + t^{-1/m}}{m^2 A\kappa}\right] e^{\pm t^{1-1/m}}$$

For the present Universe,the infinite time limit makes the above densities become zero.

It is possible to define,

$\rho_{grav} = -\frac{3H^2}{\kappa g_{00}}$ (negative energy-density of the gravitational field)

Now, let us obtain the gravitational energy of the field,

$$E_{grav} = \rho_{grav} V = -(4/3)\pi R^3 \left(\frac{3H^2}{\kappa g_{00}}\right) = -\frac{c^4 R^{4-3m}}{2Gm^2 A (mD)^{1/m-3}}$$

CONCLUSIONS

If we calculate the centripetal acceleration corresponding to the above angular speed (9.4.1), we find, for the present Universe, with $R \approx 10^{28}$cm and $c \simeq 3.10^{10}$cm./s ,

$$a_{cp} = -\omega^2 R \cong -9.10^{-8} cm/s^2. \tag{9.4.5}$$

Our model of above Section has been automatically calculated alike with (9.4.1) and (9.4.5). This value matches the observed experimentally deceleration of the NASA Pioneers' space-probes.

The solution of Section 3, is in fact a large class of solutions, for it embraces any possible deceleration parameter value,or, any power-law scale-factor.

9.5. The Fundamental Theorem for the Pioneers Anomaly Solution in General Relativity

Now we may write down the Fundamental Theorem for Pioneer Anomaly in Relativistic Cosmology. Let us rewrite the field equations of the Generalized RW´s metric, as in Section 9.1, to wit,

$$3\dot{R}^2 = \kappa(\rho + \tfrac{\Lambda}{\kappa}) g_{00} R^2 - 3k g_{00}, \tag{9.5.1}$$

and,

$$6\ddot{R} = -g_{00}\kappa\left(\rho + 3p - 2\tfrac{\Lambda}{\kappa}\right) R - 3 g_{00} \dot{R}\, \dot{g}^{00}\,. \tag{9.5.2}$$

By comparison with standard RW´s metric field equations, we write,

$$\dot{R}_t^2 = \dot{R}_\tau^2 g_{00}. \tag{9.5.3}$$

and,

$$\ddot{R}_t = \ddot{R}_\tau + \dot{R}_t \omega\,. \tag{9.5.4}$$

So, from a standard non-rotating solution in RW´s metric,you can find a corresponding rotating one, by means of the two last equations.You first obtain the square of your original scale-factor solution, and multiply by the metric time-varying coefficient, obtaining the new rotating time derivative of the scale factor squared.And, for the second time derivative of the scale-factor, you add the term $\dot{R}_t \omega = \pm cH$. Automatically, you have found the new two rotating solutions, with the exact Pioneers Anomalous deceleration,$-\omega^2 R = -c^2/R$, with angular speed $\pm\, c/R$.

9.6. The Pioneers Anomaly Solution with Universal Spin Conservation in General Relativity

Let us rewrite the field equations of the Generalized RW´s metric, as in Section 9.1, to wit,

$$3\dot{R}^2 = \kappa(\rho + \tfrac{\Lambda}{\kappa})g_{00}R^2 - 3kg_{00}, \tag{9.5.1}$$

and,

$$6\ddot{R} = -g_{00}\kappa\left(\rho + 3p - 2\tfrac{\Lambda}{\kappa}\right)R - 3g_{00}\dot{R}\,\dot{g}^{00}\,. \tag{9.5.2}$$

Consider the following solution.

$R = ct$
$\omega = \frac{c}{R} = t^{-1}$

$k = 0$
$g_{00} = t^2$
$p = \alpha\rho$
$\kappa\rho = At^{-4}$
$\Lambda = Bt^{-4}$
The conditions among the constants are,

$$A + B = 3$$
$$A(1 + 3\alpha) - 2B = 6$$

With the above solution,we find a constant spin of the Universe,namely,

$$L = MRc \propto \rho R^4 = \text{const}$$

Of course, the Pioneers deceleration is there,

$$a = -\omega^2 R = -9.10^{-36}.10^{28} = -9.10^{-8}\text{cm.s}^{-2}$$

Automatically, you have found the new two rotating solutions, with the exact Pioneers Anomalous deceleration,$-\omega^2 R = -c^2/R$, with angular speed $\pm\, c/R$.

However, with a negative angular speed, we shall not obtain spin conservation, because the function for the metric coefficient is another one.

Chapter 10

The Pioneers Anomaly and Several Relativistic Theories

10.1. On Schwarzschild's and Robertson–Walker´s Models

Local Physics extends globally with the same "laws" – if Newtonian Physics is valid locally, it is also valid globally; but, we shall show that, if General Relativity is applied locally through Schwarzschild's metric, it must be equivalent to Robertson-Walker's in the large. This should be the General Relativistic Machian picture. Lichtenegger and Mashhoon (2007) recall that the Schwarzschild´s solution, depends on boundary conditions at infinity, but the distant stars do not play the game in this solution.

Let us show that the above criterion is possible. Consider Schwarzschild's metric with a cosmological term:

$$ds^2 = g_{00}dt^2 - g_{00}^{-1}dr^2 - d\sigma^2, \tag{10.1.1}$$

where,

$$g_{00} = 1 - \frac{2GM}{c^2R} + \frac{GM\Lambda R}{3c^2}. \tag{10.1.2}$$

If we associate the Schwarzschild's g_{00} with the Machian Brans-Dicke equalities above, so that, any point of space is equivalent to any other, at each one, we have a kind of Schwarzschild's g_{00} , given by:

$$g_{00} = 1 - 2\gamma_G + 4\gamma_\Lambda = \Gamma > 0, \tag{10.1.3}$$

where Γ is a constant. A constant g_{00} , can be described as Robertson-Walker's temporal coefficient (it can be always made equal to one).

In order to give "power" to the above, we must understand that:

(a). the radial coordinate may be associated with the scale-factor in the Machian limit.
(b). we can always reparametrize the metric coordinates by making:

$$dx'^i \equiv R^2(t)dx^i \qquad , \qquad (i = 1,2,3)$$

and,

$$dt' \equiv dt.$$

We now see why Robertson-Walker's metric (at least, the flat case) is a kind of isomorphic to Schwarzschild's and, as we shall see below, its rotating analogues, in the Machian limit.

$$ds^2 = g_{00}c^2dt^2 - \frac{dr^2}{g_{00}} - d\,\Omega^2, \tag{10.1.5}$$

where,

$$d\,\Omega^2 = r^2 d\theta^2 + r^2 \sin^2\theta\; d\phi^2,$$

and,consider a circular orbit ($dr = 0$), in the equatorial plane ($d\theta = 0\ ;\ \theta = \frac{\pi}{2}$),

and, we define an angular speed, by,

$$d\phi = \omega\; dt \qquad (\omega = \text{ constant})\,, \tag{10.1.6}$$

Then, taking the extremal of the metric, which in this case is obtained by the condition,

$$\frac{ds}{dr} = 0 \tag{10.1.7}$$

we find, for the whole Universe,

$$ds = (g_{00}c^2 - \omega^2 r^2)^{1/2} dt \tag{10.1.8}$$

so that,

$$\omega \cong \frac{\beta}{R} \qquad (\beta = \text{constant}). \tag{10.1.9}$$

You can fill the same orbit $r = R$, with several "distant "stars,moving around the center.

.

We also see why Berman (2007c), has hinted that the Universe is to be considered a white (or, generally speaking, black) hole, when General Relativity with R.W.'s metric with cosmological constant, is taken into account. The fact that there is a limit where both metrics yield the same picture, has a similarity with Barrow's equivalence between Newtonian and General Relativistic Cosmologies.

CLASSICAL ORIGIN FOR THE DARK ENERGY

From Raychaudhuri's equation for a perfect fluid, we obtain the following Robertson-Walker's metric result,

$$\ddot{R} = \left[-\frac{\kappa}{6}(\rho + 3p) + \frac{1}{3}\Lambda\right] R\,. \qquad (10.1.22)$$

It is clear that the cosmological term represents the repulsive acceleration,

$$a_\Lambda = \frac{1}{3}\Lambda\, R\,. \qquad (10.1.23)$$

On the other hand, we have shown that the Universe undergoes a rotational state, so that, there is a Machian centripetal acceleration,

$$a_{cp} = -\omega^2 R\,. \qquad (10.1.24)$$

From the Machian relations of Section 8.2 and, from the result of this Section, that the angular speed depends on R^{-1} , we find that both accelerations above, depend on R , and are, from the Machian viewpoint, equivalent. Do not forget that Λ and ω^2 should depend on R^{-2} .

It does not matter if we say that the Universe rotates, or that the Universe is endowed with a negative cosmological "constant" term. We would say that the spin of the Universe stands for the Classical origin of the cosmological constant; and we need not refer to Quantum phenomena in order to generate a lambda-Universe.

CONCLUSIONS AND PREDICTIONS

The main result of this Section, depends directly on relation (10.1.3); this relation implies that local g_{00} is equal to a positive constant, when applied for the Universe, and this entails either the rotation of the Universe or a given cosmological constant, or both. The latter, implies a deep new look on cosmological theories, and in Berman (2008; 2008a,b), he showed that Robertson-Walker's metric has a hidden rotational character, along with the usual evolutionary property. The lambda-accelerating Universe was also confirmed by

recent observations.

That being the case, we predict that the left-handed Universe, is caused by rotation; global statistical analysis of rotating clusters of galaxies, and chaotic motions in the Universe, must show the left-handed property, due to rotation. All phenomena, like violation of parity and matter-antimatter asymmetry, are explained likewise. A lambda Universe means perhaps a rotating one. This makes us predict that, not only the Universe is left-handed (Barrow and Silk, 1983), but if it will be paid attention, to chaotic phenomena in the Universe, and, also to rotational states of galaxies and clusters of them, a preference for the left-hand must be found. Not only, parity violations, but also barion-antibarion asymmetries (Feynman et al., 1962) will be explained in terms of the said rotation of the Universe.

10.2. Linearized Gravitomagnetism and Pioneers Anomaly

Lichtenegger and Mashhooon (2007), while discussing Mach´s Principle, comment that Schwarzschild´s solution, intended to be "absolute", in fact depends on gravitational potentials with boundary conditions at infinity, but the distant stars do not play part of the game, ie, do not enter in the calculations.They say that Mach´s Principle according to Einstein, predicts more or less, the dragging effects.The induced rotation for arbitrarily large masses, rotating,could approach the rotation of the shell.Consider, now, that it is the whole Universe who rotates.

Berman (2007) proposed that given the fact that , from a theoretical point of view, the Universe had zero-total energy,when the gravitational interaction negative contribution was taken into account, one could estimate that the Universe should also have a spin.The numerical value of such spin, was then estimated, and pointed to an angular speed near ,

$$\omega \approx \pm \tfrac{c}{R} = 3.10^{-18} \mathrm{rad.s}^{-1}. \tag{10.2.1}$$

This formula would match the experimentally measured retardation of Pioneers space-probes, launched by NASA more than thirty years ago, called the Pioneers Anomaly (Anderson et al.,2002).Later, Berman and Berman and Gomide(2011), retrieved this numerical relation in several theoretical frameworks.Several proofs of the zero-total energy were also produced, and the least modification that Robertson-Walker metric should suffer, in order to produce the rotating scenario, was then published (Berman,2008 a).Furthermore, it was shown that the same angular speed above, followed from the rotating metric, which kept all the achievements obtained up to now by the use of RW´s metric in Standard Cosmology (Berman and Gomide, 2011).This was an exact solution of Einstein´s equations.

A rotating Universe, might however generate gravitomagnetic effects, difficult to treat with the complexities of the non-linearities of Einstein´s equations, however, it becomes treatable with the tools of linearized GRT, even if we do not write a metric.

If we take credit of the estimated angular speed above, (10.2.1), we obtain, $a \approx -\tfrac{c^2}{R} = -9.10^{-8} \mathrm{cm.s}^{-2}$. (10.2.2)

This acceleration coincides with the Pioneers anomalous deceleration.The purpose of the present Letter is to show how linearized General Relativity,while trying to get the gravitomagnetic effects,allows us to imagine a rotating Universe, coherent with Pioneers anomaly, through the gravitomagnetic formulae.

GENERAL RELATIVISTIC LINEAR PERTURBATION APPROACH

Mashhoon (2007), reviews the General Relativistic linear perturbation approach to GEM (gravitoelectromagnetism). We claim that with this linearized formalism, we can study a rotating Universe,with a local observer, acted upon by distant sources. The sources are far-away. The idea is old, it is not mine(see, for instance, Sciama,1953).While Sciama rotates the observer, I reverse the scenario and rotate the distant stars.

Far from the sources,we should have, N "stars" of mass M/N on a spherical surface, at a distance R from the point of the observer, in the "center".

Each star has a speed c and an angular momentum $L/N = RMc/N$ .In the Newtonian limit,the gravitational potentials in a pseudo-electromagnetic theory, would be

$$\Phi \approx \frac{GM}{r}, \tag{10.2.3}$$

and,

$$\vec{A} \approx \frac{G}{cr^3}\vec{L} \times \vec{r}\,. \tag{10.2.4}$$

Above, M and L represent mass and angular momentum of the source, while (10.2.3) and (10.2.4) are the potentials, such that, in the approximation,the Electromagnetic theory, is rescued, with electric and magnetic field, respectively,

$$\vec{E} = -\operatorname{grad}\Phi - \frac{\partial}{2c\partial t}\vec{A} \tag{10.2.5}$$

$$\vec{B} = \operatorname{rot}\vec{A} \tag{10.2.6}$$

To the lowest orders,the force vector is then given by,(Mashhoon,2007),

$$\vec{F} = -m\vec{E} - \frac{2m}{c}\vec{v} \times \vec{B}\,. \tag{10.2.7}$$

Notice the factor "2" in the last rhs term.It is necessary because linearized gravity is a spin-2 field.Then, the gravitomagnetic charge should be twice the gravitoelectric one.

The gravitomagnetic acceleration is,

$$a_g = \frac{2}{c}\vec{v} \times \vec{B}\,. \tag{10.2.8}$$

When we write, the radial equation , in polar coordinates,related to (10.2.7), we find,

$$\ddot{r} = -\frac{2}{c}\vec{v} \times \vec{B} + r\dot{\phi}^2 \tag{10.2.8a}$$

because each two diametrically opposed stars originate opposite contributions to the scalar potential acceleration.

Consider now a rotating Universe,and then, with Berman (2007) angular speed,

$$\omega \approx \frac{c}{R}$$

and because, from (10.2.4) and (10.2.6),

$$B \approx \frac{GL}{cR^3} \tag{10.2.9}$$

the rhs first term in (10.2.7), yield the acceleration due to gravitomagnetism, as,

$$a_g \approx -2\frac{c^2}{R}$$

with the mass obeying the well-known Brans-Dicke relation(Weinberg,1972),

$$\frac{GM}{c^2R} = \gamma \approx 1.$$

The rhs second term, however, yields, the centrifugal acceleration,

$$a_{cf} = R\omega^2 \approx \frac{c^2}{R}\,,$$

and the net result is,for the Pioneers anomaly,

$$\ddot{r} \approx -2\frac{c^2}{R} + \frac{c^2}{R} = -\frac{c^2}{R} = -9.10^{-8}\text{cm.s}^{-2}. \tag{10.2.10}$$

and, then, for the whole Universe, as expected,we find a cosmological-originated radial-field of deceleration coincident with the Pioneers anomaly.

ENERGY DENSITY OF THE ROTATING UNIVERSE

The total energy-density of the rotating Universe, will now be shown to be zero.It is composed by the relative sum of the matter energy density, and the field ´s ones.

$$\rho_{tot} = \rho_m + \rho_E + \rho_B .$$

From the electromagnetic theory,transposed by our laws of correspondence, plus relation (10.2.10),

$$\rho_E + \rho_B = -\frac{1}{4\pi G}(E^2 + 2B^2) \approx -\frac{c^4}{3\pi G R^2} , \qquad (10.2.11)$$

where the negative sign comes from the attractive properties of gravitation, and the magnetic counterpart has the factor "2" (from the charge $2m$).

Notice that, in diametrical pairs,

$$\vec{E}_{tot} = 1/2 \sum (E_i - E_i)\vec{r}/r = 0$$

but,

$$E^2 = \sum \vec{E}_i . \vec{E}_i = (\frac{GM}{R^2})^2$$

On the other hand,the inertial mass to be considered in this matter energy density, for the Universe, is M. Thus,,we write, considering (12),

$$\rho_m = \frac{Mc^2}{(4/3)\pi R^3} \approx \frac{3c^4}{4\pi G R^2} \qquad (10.2.12)$$

We find,

$$\rho_{tot} = \rho_m + \rho_E + \rho_B \approx 0 \qquad (10.2.13\ )$$

The total energy density, taken care of the negative field interaction, is zero, for the linearized perturbed,alias rotating,given metric of the Universe.This is also in agreement with the exact solution in General Relativity,by Berman and Gomide (2011).

10.3. The Pioneers Anomaly in a Variable Speed of Light Relativistic Cosmology

Berman (2007) proposed that given the fact that , from a theoretical point of view, the Universe had zero-total energy,when the gravitational interaction negative contribution was taken into account, one could estimate that the Universe should also have a spin.The numerical value of such spin, was then estimated, and pointed to an angular speed near ,

$$\pm\omega \approx \frac{c}{R} = 3.10^{-18}\text{rad.s}^{-1} . \qquad (10.3.1)$$

This formula would match the experimentally measured retardation of Pioneers space-probes, launched by NASA more than thirty years ago, called the Pioneers Anomaly (Anderson et al.,2002).Later, Berman and Berman and Gomide(2011), retrieved this numerical relation in several theoretical frameworks.Several proofs of the zero-total energy were also produced, and the least modification that Robertson-Walker metric should suffer, in order to produce the rotating scenario, was then published (Berman,2008 a). Furthermore, it was shown that the same angular speed above, followed from the rotating metric, which kept all the achievements obtained up to now by the use of RW´s metric in Standard Cosmology (Berman and Gomide, 2011). This was an exact solution of Einstein´s equations.

If we take credit of the estimated angular speed above, (10.3.1), we obtain,

$$a \approx -\frac{c^2}{R} = -9.10^{-8}\text{cm.s}^{-2}\ . \tag{10.3.2}$$

This acceleration coincides with the Pioneers anomalous deceleration. The purpose of the present, is to show how a variable speed of light,added to General Relativity, allows us to find a model of the Universe, coherent with Pioneers anomaly. A particular attempt by Shojaie(2010), leads to an approximate solution.We shall give an exact solution for the Pioneers Anomaly, with a very general model.

Rotation will not be the reason for the present solution, but it is evident that Berman and Gomide (2011) solution employs an equivalent metric.

FIELD EQUATIONS AND CONSTANT DECELERATION SOLUTION

Robertson-Walker´s flat metric, with variable c,becomes,

$$ds^2 = c^2(t)dt^2 - R^2(t)d\sigma^2\ . \tag{10.3.3}$$

The field equations,for a perfect fluid, with energy density and pressure ρ, p become,

$$H^2 = \kappa\frac{\rho}{3} \tag{10.3.4}$$

and,

$$2\frac{\ddot{R}}{R} - 2\frac{\dot{c}}{c}\frac{\dot{R}}{R} + H^2 = -\frac{\kappa}{c^2}p. \tag{10.3.5}$$

The usual Berman solution for constant deceleration parameter is,

$$R = (mDt)^{1/m} \tag{10.3.6}$$

$$c = (nBt)^{1/n} \tag{10.3.7}$$

where m, n, B and D are constants, and,

$$H = (mt)^{-1} \tag{10.3.8}$$

and the deceleration parameter q is given by,

$$q = m - 1 = -\frac{\ddot{R}R}{\dot{R}^2} \tag{10.3.9}$$.

The resultant solution is,

$$\rho = \frac{3}{\kappa m^2}t^{-2}$$

$$p = -\frac{1}{\kappa}(nB)^{2/n}\left[\frac{3}{m^2} - \frac{2}{m} - \frac{2}{mn}\right]\ t^{2/n-2}$$

PIONEERS ANOMALY SOLUTION

The observed Pioneers anomalous deceleration, coincides with,
$a = -\omega^2 R = -\frac{c^2}{R}$
where the radius of the causally related Universe is given by 10^{28}cm.

The equivalency between the generalized RW treatment and the variable speed of light, means that,

$$\omega = \pm\frac{\dot{g}_{00}}{2g_{00}} = \pm\frac{2c}{2c^2}\dot{c} = \pm(nt)^{-1}$$

We obtain exactly this relation, by specifying the conditions among constants of the above theory,

$$1 = \frac{1}{m} - \frac{1}{n}$$

$$\frac{(nB)^{2/n}}{(mD)^{2/m}} = \frac{1}{m}\left(1 - \frac{1}{m}\right)$$

The mass of the Universe, is given by,

$$M = \frac{(mD)^{3/m}}{2Gm^2}t^{(3/m)-2}$$

In the given model, a constant speed of light would ruin the calculations.On the other hand, in plain GRT,the original RW´s metric,does not lead to a solution of the above type . Of course, the formalism we have employed is somehow equivalent to the rotational solution of Berman and Gomide (2010;2011).With the abundance of arbitrary constants, we may accommodate the experimental observations of present days.In particular, for accelerating models.We could even add a lambda-term.The resultant $\ddot{R}$ may become positive.

We also find the mass obeying the well-known Brans-Dicke relation(Weinberg,1972), by adjustment of constants,

$$\frac{GM}{c^2R} = \gamma \approx 1.$$

and, then, for the whole Universe, as expected,we find a cosmological-originated field of deceleration coincident with the Pioneers anomaly.As we still have room for other requirements,we may choose equations of state, deceleration or acceleration parameters, and we may even satisfy the constant matter energy requirement.This last one,consists on taking constant Mc^2. We just require, that,

$$\frac{2}{n} + \frac{3}{m} = 2$$

We praise prior research by Shojaie and Farhoudi(2004;2004a) and Shojaie(2010).

However, there is open another alternative condition that can be obeyed,so that there will be conservation of the angular momentum or spin of the Universe,$L = MRc$ =constant.

This condition is ,

$$n = -3m/2$$

We further note that Berman and Gomide are submitting one paper with a general relativistic treatment of the Pioneers Anomaly that parallels the Machian semi-relativistic approach by Berman(2007), and another one with a full general class of solutions, not necessarily Machian,of the Pioneers Anomaly within relativistic cosmology.(Berman and Gomide,2011a;b).

10.4. Energy and Stability of Our Universe

Our main task will be to show,with simple arguments, that our possibly rotating Robertson-Walker´s Universe is a zero-total energy and stable one, in the sense that it has a reparametrized metric of Minkowski's, while the latter has been shown to be the ground state of energy level among possible universal metrics (see Witten, 1981)**. Prof. Newton da Costa collaborates here.**

THE ZERO-TOTAL ENERGY OF THE ROTATING EVOLUTIONARY UNIVERSE

The zero-total-energy of the Generalized Roberston-Walker's Universe,has been shown before, by superpotentials, and now we show by reparametrization.

Consider first Robertson-Walker's metric, added by a temporal metric coefficient which depends only on t . The line element (Gomide and Uehara,1981), becomes:

$$ds^2 = -\frac{R^2(t)}{(1+kr^2/4)^2}\left[d\sigma^2\right] + g_{00}(t)\,dt^2. \qquad (10.4.1)$$

We must mention that the idea behind Robertson-Walker's metric is the Gaussian coordinate system. Though the condition $g_{00} = 1$ is usually adopted, we must remember that, the resulting time-coordinate is meant as representing proper time. If we want to use another coordinate time, we still keep the Gaussian coordinate properties. Berman (2008a) has interpreted the generalized metric as representing a rotating evolutionary model, with angular speed given by,

$\omega = \pm\frac{\dot{g}_{00}}{2g_{00}}$

Consider the following reparametrization:

$$dx'^2 \equiv \frac{R^2(t)}{(1+kr^2/4)^2}dx^2\ , \qquad (10.4.2)$$

$$dy'^2 \equiv \frac{R^2(t)}{(1+kr^2/4)^2} dy^2 \,, \tag{10.4.3}$$

$$dz'^2 \equiv \frac{R^2(t)}{(1+kr^2/4)^2} dz^2 \,, \tag{10.4.4}$$

$$dt'^2 \equiv g_{00}(t) dt^2 . \tag{10.4.5}$$

In the new coordinates, the generalized R.W.´s metric becomes:

$$ds'^2 = dt'^2 - \left[dx'^2 + dy'^2 + dz'^2 \right] . \tag{10.4.6}$$

This is Minkowski's metric.The total energy is zero.Even in popular Science accounts (Hawking, 1996; 2001; 2003; — and Moldinow, 2010; Guth,1998), it has been generally

accepted that the Universe has zero-total energy. The first such claim, seems to be due to Feynman (1962-3). Lately, Berman (2006, 2006 a) has proved this result by means of simple arguments involving Robertson-Walker's metric for any value of the tri-curvature ($0, -1, 1$).

Berman and Gomide (2010,2011) has recently shown that the generalized Robertson-Walker's metric yielded a zero-energy pseudotensorial result. The same authors showed that the result applied in case of a rotating and expanding Universe.

Witten (1981) proved that within a semiclassical approach, Minkowski's space was in the ground state of energy, which was zero-valued. He also showed that in Classical General Relativity, this space also was the unique space of lowest energy. This last result was obtained with spinor calculus, and thus could be extended to higher dimensions whenever spinors existed. The proof was obtained through the study of the limit $h \to 0$ of a supergravity argument by Deser and Teitelboim (1977) and by Grisaru (1978), where h stands for Planck's constant.

STABILITY OF OUR UNIVERSE

The conclusion of Witten was that Minkowski's space was also stable, because perturbations in the form of gravitational waves should not decrease the total energy, because it is known that gravitational waves have positive energy. We now conclude that our Universe is also stable, due to the reparametrization above. But, first, let us deal with some conceptual issues.

We have three kinds of stability criteria:

-1) . Since a physical system shows a tendency to decay into its state of minimum energy, the criterion states that the system should not be able to collapse into a series of infinitely many possible negative levels of energy. There should be a minimum level, usually zero-valued, which is possible for the physical system.;

-2) . The matter inside the system must not be possibly created out of nothing,or else, the bodies should have positive energy.;

-3) . "Small" disturbances should not alter a state of equilibrium of the system (it tends to return to the original equilibrium state). In the case of the Universe, disturbances, of course, cannot be external.

According with our discussion, the rotating Robertson-Walker´s Universe is locally and globally stable, whenever Classical Physics is concerned. Now, Berman and Trevisan (2010), have shown that Classical General Relativity can be used to describe the scale-factor of the Universe even inside Planck´s zone, provided that we consider that the calculated scale-factor behaviour reflects an average of otherwise uncertain values, due to Quantum fluctuations.

Berman and Gomide (2010,2011) have obtained a zero-total energy proof for a rotating expanding Universe. The zero result for the spatial components of the energy-momentum-pseudotensor calculation, are equivalent to the choice of a center of Mass reference system in Newtonian theory, likewise the use of co-moving observers in Cosmology. It is with this idea in mind, that we are led to the energy calculation, yielding zero total energy, for the Universe, as an acceptable result: we are assured that we chose the correct reference system; this is a response to the criticism made by some scientists which argue that pseudotensor calculations depend on the reference system, and thus, those calculations are devoid of physical meaning.

Related conclusions should be consulted (see all Berman's references and references therein). As a bonus, we can assure that there was not an initial infinite energy density singularity, because attached to the zero-total energy conjecture, there is a zero-total energy-density result, as was pointed by Berman elsewhere (Berman, 2008). The so-called total energy density of the Universe, which appears in some textbooks, corresponds only to the non-gravitational portion, and the zero-total energy density results when we subtract from the former, the opposite potential energy density.

As Berman (2009d; f) shows, we may say that the Universe is *singularity-free* , and was created *ab-nihilo* ,;in particular, there is no zero-time infinite energy-density singularity.

Referring to rotation, it could be argued that cosmic microwave background radiation should show evidence of quadrupole asymmetry and it does not, but one could argue that the angular speed of the present Universe is too small to be detected; also, we must remark that CMBR deals with null geodesics, while Pioneers' anomaly, for instance, deals with time-like geodesics. In favor of evidence on rotation, we remark neutrinos' spin, parity violations, the asymmetry between matter and anti-matter, left-handed DNA-helices, the fact that humans and animals alike have not symmetric bodies, the same happening to molluscs. We predict that chaotic phenomena and fractals, rotations in galaxies and clusters, may provide clues on possible left handed preference through the Universe.

Berman and Trevisan (2010) have remarked that creation out-of-nothing seems to be supported by the zero-total energy calculations. Rotation was included in the derivation of the zero result by Berman and Gomide (2010). We could think that the Universes are created in pairs, the first one (ours), has negative spin and positive matter; the second member of the pair, would have negative matter and positive spin: for the ensemble of the two Universes, the total mass would always be zero; the total spin, too. The total energy (twice zeros) is also zero.

Hawking and Mlodinow (2010) conclude their book with a remark on the fact that the Universe is locally stable, but globally unstable because spontaneous creation is the reason why the Universe exists, and new creations like this may still happen. Of course, this is a question of interpretation.

We now want to make a conjecture related to the stability criteria of last Section.

A physical system is not "chaotic", if small perturbations in its initial state do not originate "large" variations in its future behaviour. According to our discussion, the Robertson-Walker´s Universe, with or without rotation, is locally and globally stable under the three criteria. As its total energy is zero, we conjecture that this type of Universe is not globally chaotic, and that the three criteria for stability imply that any such system cannot be globally chaotic altogether. We remark nevertheless, that because Einstein´s field equations are non-linear,chaos is not forbidden in a local sense.

We regret that the name of a basic result in General Relativity Theory, is called "positive energy theorem " instead of the "non-negative energy theorem".

10.5. Frame-Dragging, Mach´s Principle, and the Pioneers Anomalies

Lichtenegger and Mashhoon (2007), in a penetrating analysis of Mach´s Principle, has made some comments.First, that the distant stars do not play a role in Schwarzschild´s solution, although one needs precise boundary conditions on the gravitational potentials at infinity.I think that this is not true.The said metric, reflects a non-rotating state of the distant stars.This is implicit in the derivation of the planetary motions.Second, those authors claim that curved spacetime , appears in some solutions of the Einstein´s field equations, that are devoid of matter.We saw here, that there is an energy-density associated, for instance, with the tri-curvature of RW´s metric (see Section 7.12).Another claim, is that Go"del´s solution involves an intrinsic rotation of the matter, relative to local inertial systems.This would be anti-Machian, because the inertial properties of test masses, should be completely determined by the autonomous structer of space.Thus, the metric is not governed by the matter.This is obvious, because one may choose a metric independent of the matter field—who will forbid me of choosing an imperfect fluid, for instance, and work with any metric?That is why I suggest that Mach´s Principle is just a statement on the zero-total energy of the Universe(see Section 6.8).

Mach predicted that in Newton´s bucket experiment, the rotation of the water relative to the container, should induce centrifugal forces, though very faint, but this dragging would appear more noticeably, when measured relative to the earth´s rotation.Thirring showed the existence of Coriolis forces, inside a rotating hollow shell, and this is a frame-dragging effect.There is also a Lense-Thirring calculation, on the effects of orbital precession, around a rotating central mass, like the planets and the Sun.Orbiting satellites around the Earth should also face the dragging of the earth´s rotation around itself.

Now, I shall show a cosmological frame-dragging, in the best Machian significance.

The first Pioneers anomaly has already been tackled by me.Now, let us face the second one.The spins of the Pioneers were telemetered, and as a surprise, it shows that the on-board measurements yield a decreasing angular speed, when the space-probes were not disturbed. Turyshev and Toth (2010), published the graphs (Figures 2.16 and 2.17 in their paper), from which it is clear that there is an angular acceleration of about 0.1 RPM per three years, or,

$$\alpha \approx 1.2 \times 10^{-10}\text{rad/s}^2. \qquad (10.5.1)$$

The above is positive, because the angular speeds are taken as negative.In fact, the absolute value of the angular speeds decrease.

As the diameter of the space-probes is about 10 meters, the linear acceleration is practically the Pioneers anomalous absolute value of the deceleration , in this case, 6.10^{-8} cm.s^{-2}.The present solution of the second anomaly, confirms our first anomaly explanation.

Consider now the time derivative of the angular speed of the Universe, i.e., its angular acceleration,

$$\alpha_u = \dot{\omega} = \frac{cH}{R} = \frac{9.10^{-8}}{R} \simeq 9.10^{-36}\text{rad / s}^{-2}. \qquad (10.5.2)$$

We see that the local Pioneers second anomaly,is represented by an angular acceleration, too.

It is inversely proportional to the radii.

The above formula, can be interpreted as saying that,any local subject is affected by an angular acceleration, proportional to the inverse of the linear size,

$$\alpha = \frac{9.10^{-8}}{l} \quad \text{rad/s}^{-2} \qquad (10.5.3)$$

with (l in cm).

It can be seen that with $l = 5.10^2$cm we retrieve (10.5.1), while with $l = 10^{28}$cm, we find (10.5.2), etc.

Only for atomic dimensions this gives an important effect.

It seems that cosmological frame-dragging is, in fact, a local dragging, by the distant stars.This confirms the rotation of the Universe, and the fact that the Pioneers anomaly is a real effect and not a NASA mistake. Remember that our angular speed is negative, but, in absolute values, we have negative angular accelerations.

10.6. Gravitational Larmor Theorem and the Pioneers Anomaly

As Mashoon (2007) and Iorio (2007) have remarked, the Larmor Theorem, in Electromagnetism, established an equivalence between magnetism and rotation. The force acting on a charge q of mass m is related to the translational acceleration $-\frac{q}{m}\vec{E}$, where $\vec{E}$ is the electric field, plus the rotational frequency $\omega = \frac{q_B}{2mc}B$, where $q_B = q$ and $\vec{B}$ is the magnetic field, and the Lorentz force is expressed as:

$$\vec{F} = m\,\vec{a} = q\,\vec{E} + q_B\,\vec{v} \times \vec{B}\,. \tag{10.6.1}$$

In the gravitational counterpart, since linearized gravity is a spin-two field, different from the Maxwell's theory which is a spin-one field, we need a gravitomagnetic charge $-2m$, while the gravito-electric one is $-m$. Unfortunately, the Principle of Equivalence in General Relativity Theory, as popularly represented by Einstein's elevator, lacks the gravitomagnetic field of the source, in order that the translational acceleration of the elevator be complemented by the gravitomagnetic rotation, through the equation, for the Universe, as in Section 10.2,

$$\vec{B} = \vec{\omega}\,c = \mathrm{rot}\vec{A} \quad = c^2/R \tag{10.6.2}$$

When the gravitomagnetic potential is time-dependent, the equation of motion (10.2.7) becomes, due to equation (10.2.5), given by,

$$\vec{F} = -m\vec{E} - \frac{2m}{c}\vec{v} \times \vec{B} + \frac{2m}{c}\frac{\partial \vec{A}}{\partial t}. \tag{10.6.3}$$

For a uniform gravitomagnetic-field, we have,

$$\vec{A} = \frac{1}{2}\,\vec{B} \times \vec{R}. \tag{10.6.4}$$

so that, in the model of the rotating distant stars of Section 10.2, the total gravitomagnetic acceleration, with $v = c$, is given by,

$$\vec{a} = -\frac{2}{c}\vec{v} \times \vec{B} - \frac{1}{c}\vec{R} \times \frac{\partial \vec{B}}{\partial t}\,,$$

or,

$$a = -2B - \frac{1}{c}\dot{B}\,R = -2\omega c - R\,\dot{\omega} = -2\frac{c^2}{R} + cH \cong -\frac{c^2}{R} \cong -9 \times 10^{-8}\mathrm{cm/s^2}\,. \tag{10.6.5}$$

and, remember that,

$$B \approx \frac{GL}{cR^3}\,, \tag{10.2.9}$$

while,

$$\omega = \frac{c}{R}. \tag{10.2.10}$$

Equation (10.6.3) is the gravitational analogue of the Larmor Theorem for the case when the gravitomagnetic field is time-dependent. We found the correct Pioneers deceleration, with a positive angular speed.

Alternatively, we could have defined $\omega = -c/R$, and $B = -\omega c$, and found the same results.

More details can be found in Iorio's excellent book (Iorio, 2007). Brans-Dicke relation was used above, several times.The time-dependent magnetic contribution to the total acceleration, cH ,is due to the centrifugal term in Section 10.2.

Chapter 11

Possible Solutions to the Pioneers Anomaly

11.1. Evidence on the Rotation of the Universe

Birch (1982;1983) has been one of the first researchers who reported possible Universal Rotation data.

Now let us see some of the recent experimental evidence on the rotation or vorticity in the Universe.

.....1st.) Su and Chu (2009) obtained, for a particular current model of the present Universe,a superior limit $\omega \leq 4.10^{-17}$rad.s^{-1}.

.....2nd.) Chechin (2010), considering cosmic vacuum,and the global rotation compared with the induced rotation of elliptical galaxies,estimates $\omega \sim 10^{-19}$rad.s^{-1}.

.....3rd.)With the data for rotation of the polarization of CMBR, which points to an angle of 0.1 rad (see Ni,2008), we find, dividing by the age of the Universe, $\omega \sim 10^{-19}$rad.s^{-1}.

We conclude that the rotation of the Universe, is real, and is the natural explanation for the Pioneers Anomaly.The fact that the sign of the angular speed could be negative, and not positive,makes the General Relativistic theory explain the left-handed preference. The spinning down of the spacecrafts could also be explained by us, through rotation of the Universe, and then, there is an evidence of a cosmological frame-dragging.

11.2. Thermal Emission as Deceleration Cause

Rievers and Lamerzahl (2011) have reported that they could explain the anomaly by means of a high precision thermal modelling, for complex systems,but, however,an engineer consulted by me says that the risk with complex systems is that modelling correctly is difficult for complex systems, and there is no means of being sure that cause and effect are in fact covered rigorously. He comments that it is almost as difficult a task, as predicting human behavior by making a model of one person in the computer, forty years after his death.

Worse, why do not elliptical orbiters show the same thermal emission problems????

11.3. MOND (Modified Newtonian Dynamics) by Milgrom

The Newtonian modified law by M. Milgrom(1983), could eventually explain a deceleration, but it is a heuristic law. Beckenstein (2004) included this law as a limit of a general gravitational theory. Very interesting indeed.

11.4. Other Explanations

Honestly speaking, I doubt that there is other viable explanation, besides the rotation of the Universe, or its equivalent variable speed of light theory.

11.5. Concluding Remarks

We have seen that General Relativity accounts for a possible rotating Universe, and the amount of the deceleration coincides with the P.A.We have shown that even variable speed of light theory or Machian semi-relativistic theories also lead to the P.A.There is a point that needs clarification. According to Raychaudhuri´s equation,if we consider a non-shearing case,a non-accelerated system would be described by the equation, adapted to RWs original metric,

$$6\ddot{R} = -\kappa\left(\rho + 3p - 2\tfrac{\Lambda}{\kappa}\right)R + 4\omega^2 R\,, \tag{11.5.1}$$

while, in the Generalized RW´s metric,

$$6\ddot{R} = -g_{00}\kappa\left(\rho + 3p - 2\tfrac{\Lambda}{\kappa}\right)R - 3g_{00}\dot{R}\,\dot{g}^{00} = -g_{00}\kappa\left(\rho + 3p - 2\tfrac{\Lambda}{\kappa}\right)R + 6\dot{R}\omega\,, \tag{11.5.2}$$

but for the Generalized RW´s metric, there are two different solutions,and we would take the negative angular speed solution, in order to account for a left-handed Universe. In our semi-relativistic treatment of Chapter 8, what really was needed was a solution for L^2,so that one could choose, if necessary in order to coincide with Chapter 9, a negative angular speed.In Chapter 9,the angular speed can also be chosen with a negative sign; this can be done, for all that matters is the centripetal acceleration, which depends on the square of ω. However, we must remeber that a positive $\omega = c/R$ also rotates the Universe with the Pioneers deceleration.

We conclude that the Raychaudhuri´s vorticity is NOT what we call here the angular speed of the rotation of the Universe.What we have shown, is the rotation of the entire spatial Universe around the orthogonal time-axis.

By increasing sophistication, we may develop scientific theories which could, *a priori*, cover almost any possible characteristic of the Universe, that could eventually be observed. By the same token, the reader should remember that current theoretical cosmological models may easily be turned down by future astronomical observations. However, at the same time, scientists would come with another "many" ones, that could be adapted to "new" astronomical data, which on its own, could afterwards go also to the *oblivium* .The Pioneers Anomaly seems to obey the known laws of Physics, if the Universe rotates, or the speed of light is variable according to our model. The secondary anomaly, the spinning down of the spacecraft, received an explanation as due to the rotation of the Universe. The left-hand of creation is also accounted by equation (11.5.2) with a negative angular speed. This introduces a partial decelerating contribution, but, if lambda is large enough, and positive, in order to produce a larger acceleration, $\ddot{R} \geq 0$, and the Pioneers anomalies will still be there.

That is the way Science is developed. We see that Mathematics, Physics, and reality, work together blending into a common framework.

Part V

Bibliography and References

Bibliography and References

Abbot, L.F. (1985) - *Physics Letters B* **249**, 332.

Abdel-Rahman,A.-M.M. (1992) - *Physical Review D* **45**, 3497.

Abramo, L.R.W.; Brandenberger, R.H.;Mukhanov, V.F. (1997) - *Physical Review D* **56**, 3248.

Adler, R.; Bazin, M,; Schiffer, M. (1975) - *Introduction to General Relativity* - 2nd. edtn., McGraw-Hill, N.Y.

Adler, R.J.; Silbergleit, A.S. (2000) - *International Journal of Theoretical Physics*, **39**, 1291.

Aguirregabiria, J.M. et al. (1996) - *Gen. Rel. and Grav.* **28**, 1393.

Albrecht, A.; Magueijo, J.(1998) *A time varying speed of light as a solution to cosmological puzzles*-preprint.

Anderson, J.D. et al. (1999) - PRL **83**, 1891.

Anderson, J.D. et al. (2002) - PRD, **65**, 082004.

Arbab, A.I.(1997) - GRG **29**, 61.

Arbab, A.I.; Abdel-Rahman, A.-M.M. (1994) - Physical Review D **50**, 7725.

Augustine, Saint (1958) - *City of God*, Spanish Edition , BAC, Madrid .(See L. XI, chapter 5).

Bahcall, J.N. ; Schmidt, M. (1967) - Physical Review Letters **19**, 1294.

Bandyopadhyay, N. (1977). *J. Phys.* **A10**, 189.

Banerji, S. (1974) Physical Review, **D9**, 877.

Banks, T. (1984) - *Physical Review Letters* **52**, 1461.

Banks, T. (1988) - *Nuclear Physics B* **309**, 493.

Baptista, J.P. et al. (1986). Rev. Bras. Fis., **16**, 257.

Barr, S.M.(1987) - *Physical Review D* **36**, 1691.

Barrow, J.D. (1983). in *The Very Early Universe.*, Gibbons, G.N., Hawking, S.W., and Siklos, S.T.C., eds., Cambridge University Press, Cambridge.

Barrow, J.D. (1987) Phys. Rev., **D35**, 1805.

Barrow, J.D. (1990) - in *Modern Cosmology in Retrospect.* Edited by B.Bertotti, R. Balbinot, S.Bergia and A.Messina, Cambridge U.P., Cambridge.

Barrow, J.D. (1990 a) *Phys Lett.***B235**,40.

Barrow, J.D. (1991) *Theories of Everything*, Oxford UP, Oxford.

Barrow, J.D. (1992) *Phys.Rev* **D46**, R3227.

Barrow, J.D. (1993) *Phys. Rev* **D47**, 5329.

Barrow, J.D. (1993 a) *Phys.Rev.* **D48**, 3592.

Barrow, J.D. (1994) *The Origin of the Universe, Basic Books*, NY.

Barrow, J.D. (1995) Phys. Rev **D51**, 2729.

Barrow, J.D. (1996) MNRAS **282**, 1397.

Barrow, J.D. (1997) *Varying G and Other Constants* , Los Alamos Archives http://arxiv.org/abs/gr-qc/9711084 v1 27/nov/1997.

Barrow, J.D. (1998) *Cosmologies with Varying Light Speed*, Los Alamos Archives http://arxiv.org/abs/Astro–Ph 9811022 v1.

Barrow, J.D. (2002) - Los Alamos Archives http://arxiv.org/abs/gr-qc/0211074

Barrow, J.D. (2002) - Los Alamos Archives http://arxiv.org/abs/gr-qc/0209080

Barrow, J.D. (2004) - Los Alamos Archives http://arxiv.org/abs/gr-qc/0409062

Barrow, J.D. (2004) - Los Alamos Archives http://arxiv.org/abs/gr-qc/0403084

Barrow, J.D. (2005) - Los Alamos Archives http://arxiv.org/abs/astro-ph/0503434

Barrow, J.D. ; Papers in http://arXiv.org not cited above

Barrow, J.D. ; et al (2003) - Los Alamos Archives http://arxiv.org/abs/astro-ph/0307227

Barrow, J.D. ; et al (2003) - Los Alamos Archives http://arxiv.org/abs/gr-qc/0305075

Barrow, J.D. ; et al (2003) - Los Alamos Archives http://arxiv.org/abs/astro-ph/0303014

Barrow, J.D. ; et al (2003) - Los Alamos Archives http://arxiv.org/abs/gr-qc/0302094

Barrow, J.D. ; et al (2004) - Los Alamos Archives http://arxiv.org/abs/astro-ph/0406369

Barrow, J.D.;Carr, B.J. (1996) Phys. Rev **D54**, 3920.

Barrow, J.D.;Cotsakis S. (1988) Physics Letters **B214** 515.

Barrow, J.D.;Daborwski, M.P.(1995) – MNRAS **275**, 850.

Barrow, J.D.;Gotz, G. (1989) Class.Quantum Gravity **6**, 1253.

Barrow, J.D. ; Hervik, Sigbjorn (2002) - Los Alamos Archives http://arxiv.org/abs/gr-qc/0206061

Barrow, J.D. ; Hervik, Sigbjorn (2002) - Los Alamos Archives http://arxiv.org/abs/gr-qc/0302076

Barrow, J.D. ; Hervik, Sigbjorn (2003) - Los Alamos Archives http://arxiv.org/abs/gr-qc/0304050

Barrow, J.D. ; Levin, Janna (2003) - Los Alamos Archives http://arxiv.org/abs/gr-qc/0304038

Barrow, J.D.;Liddle, A.R.(1993) Phys. Rev **D47**, R5219.

Barrow, J.D.;Maeda, K.(1990) Nucl. Phys. **B341**, 294.

Barrow, J.D.;Magueijo J.(1999)– *Solving the Flatness and Quasi Flatness Problems in Brans-Dicke Cosmologies with a Varying Light Speed* , Los Alamos Archives http://arxiv.org/abs/astro-ph/9901049 5/01/1999.

Barrow, J.D. ; Magueijo, J. (2005) - Los Alamos Archives http://arxiv.org/abs/astro-ph/0503222

Barrow, J.D.;Mimoso, J.P.(1994) Phys. Rev **D50**, 3746.

Barrow, J.D.;Mimoso, J.P; Maia, M.R.G.(1993) Phys. Rev **D48**, 3630.

Barrow, J.D. ; Mota, D. F. (2002) - Los Alamos Archives http://arxiv.org/abs/gr-qc/0207012

Barrow, J.D. ; Mota, D. F. (2002) - Los Alamos Archives http://arxiv.org/abs/gr-qc/0212032

Barrow, J.D.;Saich, P.(1990) Phys. Lett **B249**, 406.

Barrow, J.D. ; Scherrer, Robert, J.(2004) - Los Alamos Archives http://arxiv.org/abs/astro-ph/0406088

Barrow, J.D. ; Shaw, Douglas, J. (2004) - Los Alamos Archives http://arxiv.org/abs/gr-qc/0412135

Barrow, J.D. ; Subramanian, Kandaswamy (2002) - Los Alamos Archives http://arxiv.org/abs/astro-ph/0205312

Barrow, J.D. ; Tsagas, Christos, G.(2003) - Los Alamos Archives http://arxiv.org/abs/gr-qc/0309030

Barrow, J.D. ; Tsagas, Christos, G.(2003) - Los Alamos Archives http://arxiv.org/abs/gr-qc/0308067

Barrow, J.D. ; Tsagas, Christos, G.(2004) - Los Alamos Archives http://arxiv.org/abs/gr-qc/0411070

Barrow, J.D. ; Tsagas, Christos, G.(2004) - Los Alamos Archives http://arxiv.org/abs/gr-qc/0411045

Barrow, J.D.; Tipler, F.J. (1996) *The Anthropic Cosmological Principle* ,Oxford University Press, Oxford.

Beesham, A. (1986) - *International Journal of Theoretical Physics* **25**, 1295.

Beesham, A. (1993) - *Physical Review D* **48**, 3539.

Beesham, A. (1995) - GRG **27**, 15.

Bekenstein, J.D. (1974) - *Ann. Phys. (New York)* **82**, 535.

Bekenstein, J.D. (2004) - *PRD* **70**, 083509.

Bergmann, P.G. (1968) Int.J.Theor.Phys. **1**, 25.

Berman, M.S. (1981) M.Sc. Thesis. It can be found online trhough the site www.sophia.bibl.ita.br/biblioteca/index.html (insert keyword "Einstein" and user-name "Berman".)

Berman, M.S. (1983) - *Nuovo Cimento* **74B**, 182.

Berman, M.S. (1988) - GRG **20**, 841.

Berman, M.S. (1988 b) - GRG **21**, 967.

Berman, M.S. (1989) - GRG **21**, 967.

Berman, M.S. (1989) - *Physics Letters A* **142**, 335.

Berman, M.S. (1989 b) - Physics Letters **A 139**, 119.

Berman, M.S. (1989 d) - Physics Letters **A 142**, 227.

Berman, M.S. (1990) - *International Journal of Theoretical Physics* **29**, 1419.

Berman, M.S. (1990 a) - *International Journal of Theoretical Physics,* **29**, 567-570.

Berman, M.S. (1990 b) - GRG **22,**389.

Berman, M.S. (1990 c) - International Journal of Theoretical Physics **29**, 1415.

Berman, M.S. (1990 e) - International Journal of Theoretical Physics **29**, 571.

Berman, M.S. (1990 f) - Nuovo Cimento **B 105**, 1373.

Berman, M.S. (1990 g) - Nuovo Cimento **B 105**, 239.

Berman, M.S. (1990 h) - Nuovo Cimento **B 105**, 235.

Berman, M.S. (1990 i) - International Journal of Theoretical Physics **29**, 1423.

Berman, M.S. (1991) - *GRG* **23**, 465.

Berman, M.S. (1991 b) - *Physical Review D* **43**, 1075.

Berman, M.S. (1991 c) - *GRG* **23**, 1083.

Berman, M.S. (1992) - *International Journal of Theoretical Physics* **31**, 1447.

Berman, M.S. (1992 b) - *International Journal of Theoretical Physics* **31**, 321.

Berman, M.S. (1992 c) - *International Journal of Theoretical Physics* **31**, 1217.

Berman, M.S. (1992 d) - *International Journal of Theoretical Physics* **31**, 329.

Berman, M.S. (1992 e) - *International Journal of Theoretical Physics* **31**, 1451.

Berman, M. S. (1994) - *Astrophys. Space Science,* **222**, 235.

Berman, M.S. (1994 a) - *International Journal of Theoretical Physics* **33**, 1929.

Berman, M.S. (1994 c) - *Astrophysics and Space Science* **215**, 135. (by mistake, this journal published again Berman, 1992 c).

Berman, M.S. (1996) - International Journal of Theoretical Physics **35**, 1789.

Berman, M.S. (1996 a) - International Journal of Theoretical Physics **35**, 1033.

Berman, M.S. (1996 b) - International Journal of Theoretical Physics **35**, 1719.

Berman, M.S. (1997) - International Journal of Theoretical Physics **36**, 1249.

Berman, M.S. (1997 a) - International Journal of Theoretical Physics **36**, 1461.

Berman, M.S. (2004) "Energy of Kerr-Newman Black-Holes and Gravitomagnetism" – Los Alamos Archives, http://arxiv.org/abs/gr-qc/0407026.

Berman, M.S. (2004a) ."Brief History of Black-Holes" Los Alamos Archives, http://arxiv.org/abs/gr-qc/0412054

Berman, M.S. (2005) - Los Alamos Archives http://arxiv.org/abs/physics/0507110

Berman, M.S. (2006) - *Energy of Black-Holes and Hawking's Universe.* In *Trends in Black-Hole Research,* Chapter 5. Edited by Paul Kreitler, Nova Science, New York.

Berman, M.S. (2006 b) - *Energy, Brief History of Black-Holes, and Hawking's Universe.* In *New Developments in Black-Hole Research,* Chapter 5. Edited by Paul Kreitler, Nova Science, New York.

Berman, M.S. (2006 c) - *On the Machian Properties of the Universe*, submitted to publication. Los Alamos archives: http://www.arxiv.org/abs/physics/0610003

Berman, M.S. (2006 d) - *On the Magnetic Field and Entropy Increase of a Machian Universe* - submitted to publication. Los Alamos Archives http://arxiv.org/abs/physics/0611007.

Berman, M.S. (2006 e) - *The Pioneer anomaly and a Machian Universe* - submitted to publication. Los Alamos Archives http://arxiv.org/abs/physics/0606117.

Berman, M.S. (2006 f) - Los Alamos Archives http://arxiv.org/abs/gr-qc/0605063

Berman, M.S. (2006 g) - Los Alamos Archives http://arxiv.org/abs/gr-qc/0605092

Berman, M.S. (2006 h) - Los Alamos Archives http://arxiv.org/abs/physics/0612007

to be published by Astrophysics and Space Science.

Berman, M.S. (2006 i) - Los Alamos Archives http://arxiv.org/abs/physics/0607005

Berman, M.S. (2006 j) - Los Alamos Archives http://arxiv.org/abs/physics/0608053

Berman, M.S. (2006 k) - Los Alamos Archives http://arxiv.org/abs/physics/0606208

Berman, M.S. (2006 l) - Los Alamos Archives http://arxiv.org/abs/physics/0609026

Berman, M.S. (2007) - *Introduction to General Relativity and the Cosmological Constant Problem.* Nova Science, New York.

Berman, M.S. (2008) - *A Primer in Black Holes, Mach´s Principle, and Gravitational Energy.* Nova Science, New York.

Berman, M.S. (2007 c) - Are Mass and Length Quantized, submitted. See also, Los Alamos Archives, http://arxiv.org/abs/physics/0707.3290[physics.genph]

Berman, M.S. (2007 d) "Misconceptions in Halliday,Resnick, and Walker's textbook" Los Alamos Archives, http://arxiv.org/abs/physics/0507110 New submitted version:Misconceptions on Relativity, Gravitation and Cosmology.(2008).

Berman, M.S. (2007 e) - *Introduction to General Relativistic and Scalar-Tensor Cosmologies.* Nova Science Publishers, New York.

Berman, M.S. (2006) "Cosmological Model For the Very Early Universe in B.D. Theory" — Los Alamos Archives, http://arxiv.org/abs/gr-qc/0605092

Berman, M.S. (2006) "On the Rotational and Machian Properties of the Universe" Los Alamos Archives, http://arxiv.org/abs/physics/0610003 Submitted. [former title: "On the Machian Properties of the Universe"].

Berman, M.S; Gomide, F.M. (2006) - *On a time variation of Neutrinos' mass* - submitted. See also Los Alamos archives: http://www.arxiv.org/abs/physics/0606208.

Berman, M.S. (2008) ."Shear and Vorticity in a Combined Einstein-Cartan-Brans-Dicke Inflationary Lambda-Universe"—Letter, Astrophys. Space Science 314,79-82 . For a previous version, see:

Los Alamos Archives,http://arxiv.org/abs/physics/0607005.

Berman, M.S. (2007) ."Gravitomagnetism and Angular Momenta of Black-Holes". .Revista Mex. Astron. Astrofís.43,297-301 . Los Alamos Archives,http://arxiv.org/abs/physics/0608053.

Berman, M.S. (2006) "On the Magnetic Field, and Entropy Increase, in a Machian Universe".Main result published within Berman,M.S. (2007) "The Pioneer Anomaly and a Machian Universe"-

Astrophys. Space Science 312,275 . Los Alamos Archives,http://arxiv.org/abs/physics/0606117. Los Alamos Archives,http://arxiv.org/abs/physics/0611007.

Berman, M.S. (2006) - "On the Rotational and Machian Properties of the Universe" Los Alamos Archives, http://arxiv.org/abs/physics/0610003 . Submitted. [former title: "On the Machian Properties of the Universe"].

Berman, M.S. (2007) - "Is the Universe a White-hole?".Astrophys. Space Science, 311 , 359-361. Los Alamos Archives, http://arxiv.org/abs/physics/0612007

Berman, M.S. (2007) - "Shear and Vorticity in Inflationary Brans-Dicke Cosmology with Lambda-term"- Astrophysics and Space Science 310,205.Los Alamos Archives, http://arxiv.org/abs/physics/0703244

Berman, M.S. (2007) - "The Pioneer Anomaly and a Machian Universe"- Astrophys. Space Science 312,275 (2007) . Los Alamos Archives,http://arxiv.org/abs/physics/0606117.

Berman, M.S. (2007) - "Simple derivation of Schwarzschild, Lense- Thirring,Reissner-Nordström,Kerr, and Kerr-Newman metrics" Los Alamos Archives http://arxiv.org/abs/physics/0702014

Berman, M.S. (2007) - "Are Mass and Length Quantized?" Submitted under new title and in extended version "Cosmic and Micro-Cosmic Scales of Length, Mass, and angularity". For early version, see Los Alamos Archives http://arxiv.org/abs/0707.3290

Berman, M.S. (2007) - ."Gravitomagnetism and Angular Momenta of Black-Holes". .Revista Mex. Astron. Astrofís.43,297-301. Los Alamos Archives, http://arxiv.org/abs/physics/0608053.

Berman, M.S. (2008) - "Misconceptions in Halliday,Resnick, and Walker's textbook" Los Alamos Archives, http://arxiv.org/abs/physics/0507110 New submitted version: Misconceptions on Relativity, Gravitation and Cosmology.

Berman, M.S. (2008) - ”Shear and Vorticity in a Combined Einstein-Cartan-Brans- Dicke Inflationary Lambda-Universe”—Letter, Astrophys. Space Science 314,79-82 . For a previous version, see: Los Alamos Archives, http://arxiv.org/abs/physics/0607005.

Berman, M.S. (2008) - “On a time variation of Neutrino´s Mass” Los Alamos Archives, http://arxiv.org/abs/physics/0606208. New version submitted with title ”The Pioneer Anomaly and a Machian Universe-Part II”

Berman, M.S. (2008) - ”On the Machian Origin of Inertia”(2008).Letter.Astrophysics Space Science, 318,269-272 (2008) Preliminary version, Los Alamos Archives, http://arxiv.org/abs/physics/0609026

Berman, M.S. (2008) - “Energy and Angular Momentum of Dilaton Black-Holes” Revista Mex. Astron. Astrof. 44,285-291. Los Alamos Archives http://arxiv.org/abs/0804.0881

Berman, M.S. (2008) - .”A General Relativistic Rotating Evolutionary Universe”- Astrophys. Space Science 314,319-321 . Preliminary version posted in Los Alamos Archives http://arxiv.org/abs/0712.0821

Berman, M.S. (2008) - ”A General Relativistic Rotating Evolutionary Universe Part II”, Astrophys. Space Science,315,367-369 (2008). Posted with another title, in Los Alamos Archives http://arxiv.org/abs/0801.1954

Berman, M.S. (2008) - ”General Relativistic Machian Universe”-Letter.Astrophys. & Space Science 318,273-277. Los Alamos Archives http://arxiv.org/abs/0803.0139

Berman, M.S. (2008) - ”Shear and Vorticity in an Accelerating Brans-Dicke Universe with Torsion”. Astrophys. Space Science, 317, 279-281 (2008). Los Alamos Archives, http://arxiv.org/abs/0807.1866

Berman, M.S. (2009) - “Lambda-Universe in Scalar-Tensor Gravity”-Astrophysics and Space Science, 323, 103.

Berman, M.S. (2009) - “On the zero-energy Universe” -Intl.Journ. Theor. Phys.,48,3278,(2009).DOI 10.1007/s10773-009-0125-8. Los Alamos Archives, http://arxiv.org/abs/gr-qc/0605063

Berman, M.S. (2009) - ”Pryce-Hoyle Tensor in a Combined Einstein-Cartan-Brans- Dicke Model”-International Journ. Theoret. Physics, 48,836 . Preliminary version, at Los Alamos Archives, http://arxiv.org/abs/0805.0448

Berman, M.S. (2009) - ”Simple Model with time-varying fine-structure “constant”-RevMexAA, 45,139-142,(2009).

Berman, M.S. (2009) - “Gravitons,Dark Matter, and Classical Gravitation” AIP Conference Proceedings (Eds): T. E. Simos, G. Psihoyios, and C.. Tsitouras, 1168,.1064-1104. Los Alamos Archives http://arxiv.org/abs/0806.1766

Berman, M.S. (2009) - "Entropy of the Universe" – International Journal Theoret. Phys.,48,1933,(2009). Los Alamos Archives http://arxiv.org/abs/0904.3135

Berman, M.S. (2009) - "Entropy of the Universe and Standard Cosmology" , International Journal Theoret. Physics 48, 2286 .

Berman, M.S. (2009) - "General Relativistic Singularity-Free Cosmological Model"- Astrophys. & Space Science,321,157. Los Alamos Archives http://arxiv.org/abs/0904.3141.

Berman, M.S. (2009) - "Why the initial Infinite Singularity of the Universe is not there" Internatl.Journ.Theoret. Phys.48,2253 . Los Alamos Archives http://arxiv.org/abs/0904.3143.

Berman, M.S. (2009) - "On the Newtonian and General Relativistic Energies of a Rotating Body"-Astrophys. and Space Science, 324,1.

Berman, M.S. (2009) - Static Scalar-Tensor Universes and Gravitational Waves Ámplification(2009)- Astrophys and Space Science,323,99.

Berman, M.S. (2009) - "On Sciama´s Machian Cosmology"-Interntl. Journ. Theor.Phys.,48,3257, DOI 10.1007/s10773-009-0112-0

Berman, M.S. (2010) - "Simple Model with time-varying fine-structure "constant"- Part II" - RevMexAA, 46,23-28. Preliminary version, Los Alamos Archives, http://arxiv.org/abs/0809.4793

Berman, M.S. (2010) - "Amplification of Gravitational Waves in Scalar-Tensor Accelerating Lambda-Universes", Astrophysics and Space Science, 325, 283-286 . DOI 10.1007/s10509-009-0180-x http://www.springerlink.com/content/w24nr73u74777h7t/

Berman, M.S. (2010) – "Amplification of Gravitational Waves in Scalar-Tensor Inflationary Lambda-Model"-Interntl. Journ. of Theoret. Phys., 49, 232-237. DOI 10.1007/s10773-009-0161-4.

Berman, M.S. – Gravitomagnetism, Universal Rotation and Pioneers Anomaly – submitted

Berman, M.S. – General Relativity with Variable Speed of Light and Pioneers Anomaly – submitted

Berman, M.S. – Cosmology and the Two Pioneers Anomalies – submitted

Berman, M.S. – The Exact Brans-Dicke Relation, with Variable Speed of Light and Pioneers Anomaly – submitted.

Berman, M.S.; Costa, N.C.A. da (2010) – "On the stability of our Universe"- with N.C.A. da Costa. Cornell University Library, http://arxiv.org/PS_cache/arxiv/pdf/1012/1012.4160v1.pdf

Berman, M.S.; Gomide, F.M. (1987) - *Tensor Calculus and General Relativity: an introduction.* (in Portuguese), McGraw-Hill, São Paulo, 2nd ed (first edition in 1986).

Berman, M.S; Gomide, F.M. (1988) - *GRG* **20**, 191.

Berman, M.S; Gomide, F.M. (1994) - *International Journal of Theoretical Physics* **33**, 1931.

Berman, M.S.; Gomide, F.M. (2010) - "General Relativistic Treatment of the Pioneers Anomaly", - submitted . Los Alamos Archives,arxiv:1011.4627 v2 [physics.gen-ph].Co-author-F.M.Gomide

Berman, M.S.; Gomide, F.M. (2011) - Chapter 12 of *The Big-Bang: Theory, Assumptions and Problems,* ed. by O'Connel and Hale, Nova Science Publishers, New York. (at press)

Berman, M.S.; Gomide, F.M.; – Relativistic Cosmology and the Pioneers Anomaly - submitted

Berman, M.S.; Marinho Jr., R. M. (1996) - Letter to the Editor, *Physics Today,* **49,** 13.

Berman, M.S.; Marinho Jr., R. M. (1996 b) - Nuovo Cimento B, **111**, 1279.

Berman, M.S.; Marinho Jr., R. M. (2001) - *Astroph. Space Science* **278**, 367.

Berman, M.S.; Paim, T. (1990) - Nuovo Cimento B, **105**, 1377.

Berman, M.S.; Som, M.M. (1989) - *Physics Letters A* **142**, 338.

Berman, M.S; Som, M.M. (1989 b)Progress Theor.Phys, **81**, 823.

Berman, M.S; Som, M.M. (1989 c)Phys. Letters **A 136**, 206.

Berman, M.S.;Som, M.M. (1989 d). Nuovo Cimento **103B,N.2**, 203.

Berman, M.S.; Som, M.M. (1989 e) - Physics Letters **A 136**, 428.

Berman, M.S.; Som, M.M. (1989 f) - Physics Letters **A 139**, 119.

Berman, M.S.; Som, M.M. (1989 g) - GRG , **21**, 967-970.

Berman, M.S; Som, M.M. (1990) - *GRG* **22**, 625.

Berman, M.S.; Som, M.M. (1990 b) - *International Journal of Theoretical Physics* **29**, 1411.

Berman, M.S; Som, M.M. (1992) - International Journal of Theoretical Physics **31**, 325.

Berman, M.S.; Som, M.M. (1993) - Astrophysics and Space Science, **207**, 105.

Berman, M.S.; Som, M.M. (1993 b) - Journal of Mathematical Physics, **34(1)**, 111 .

Berman, M.S; Som, M.M (1995) - Astrophys. Space.Sci, **225**, 237.

Berman, M.S., Som, M.M. (2007) – Natural Entropy Production in an inflationary model for a Polarized Vacuum, Astrophysics and Space Science, 310, 277 (2007). See also Los Alamos Archives, http://arxiv.org/abs/physics/0701.070

Berman, M.S.; Som, M.M. (2007) - "Natural Entropy Production in an Inflationary Model for a Polarized Vacuum". Astrophys.and Space Scie,310,277. Los Alamos Archives, http://arxiv.org/abs/physics/0701070. DOI: 10.1007/s10509-007-9514-8 Springuer Link http://www.springerlink.com/content/u15638x8v887h450/

Berman,M.S; Som, M.M.; Gomide, F.M. (1989) - GRG **21**, 287.

Berman, M.S.; Trevisan, L.A. (2001) -On the Creation of the Universe out of Nothing- Los Alamos Archives http://arxiv.org/abs/gr-qc/0104060

Berman,M.S.; Trevisan, L.A. (2001a) - Possible Cosmological Implications of Time Varying Fine Structure Constant-Los Alamos Archives http://arxiv.org/abs/gr-qc/0112011

Berman,M.S.; Trevisan, L.A. (2001 b) -On a time varying fine structure constant - Los Alamos Archives http://arxiv.org/abs/gr-qc/0111102

Berman,M.S.; Trevisan, L.A. (2001 c) - Estimate on the deceleration parameter in a Universe with variable fine structure constant-Los Alamos Archives http://arxiv.org/abs/gr-qc/0111101

Berman,M.S.; Trevisan, L.A. (2001 d) -Inflationary phase in Generalized Brans-Dicke theory"-Interntl. Journal Theor. Phys. **48**,1929 . Los Alamos Archives http://arxiv.org/abs/gr-qc/0111098

Berman,M.S.; Trevisan, L.A. (2001 e) -Static Generalized Brans-Dicke Universe and Gravitational Waves Amplification- Los Alamos Archives http://arxiv.org/abs/gr-qc/0111099

Berman,M.S.; Trevisan, L.A. (2001 f) - Amplification of Gravitational Waves During Inflation in Brans-Dicke Theory -Los Alamos Archives http://arxiv.org/abs/gr-qc/0111100

Berman,M.S.; Trevisan, L.A. (2002) - Inflationary Phase with Time Varying Fundamental Constants - Los Alamos Archives http://arxiv.org/abs/gr-qc/0207051

Berman, M.S.; Trevisan, L.A. (2009) - "Inflationary phase in Generalized Brans-Dicke theory"(2009)- Interntl. Journal Theor. Phys. 48,1929 – Los Alamos Archives, http://arxiv.org/abs/gr-qc/0111098

Berman, M.S.; Trevisan, L.A. (2010) - "On the creation of the Universe out of nothing"- Intern. Journ. Modern Phys.D19,1309-1313. DOI 10.1142/s0218271810017342 For old version (year 2001) see Los Alamos Archives, http://arxiv.org/abs/gr-qc/0104060 This paper was included as part of Chapter 9 of book listed above as 3.h.,and Chapters 05 of 3.f. and 3.g. also listed above.

Bernardis, P. et al. (2000), *Nature* **404**, 955;

Berry, M.V. (1989) - *Principles of Cosmology and Gravitation,* Adam Hilger, Bristol.

Bertolami, O.(1986) - *Nuovo Cimento B* **93**, 36.

Bertolami, O. (1986 b). Fortschritte der Physik, **34**, (12), 829.

Birch, P. (1982) - Nature, **298**, 451.

Birch, P. (1983) - Nature, **301**, 736.

Bondi, H.; Gold, T. (1948). MNRAS, **108**, 252.

Bonnor, W.; Cooperstock, F.I. (1989) - *Phys. Lett. A* **139**, 442.

Boyer, R.H.; Lindquist, R.W. (1967) - Journal of Mathematical Physics, **8**, 265.

Brans, C.; Dicke, R. H. (1961) - *Phys. Review* **124**,925.

Brown, J.D.; Teitelboim, C.(1987) - *Physics Letters B* **195**, 177.

Burd A.B., Barrow J.D. (1988) Nucl. Physics, **B308**, 929.

Carmeli, M. (1982) - *Classical Fields* - Wiley, N.Y.

Carroll, S. M. (2001), *Living Rev. Rel.* **4**, 1.

Cartan, E. (1986) - *On manifolds with an affine connection and the Theory of General Relativity,* Bibliopolis, Firenze.

Carvalho, J.C.; Lima, J.A.S. ;Waga, I. (1992) - *Physical Review D* **46**, 2404.

Chaliassos, E. (1987). Physica, **144A**, 390.

Chechin, L.M. (2010) - Astron. Rep. **54**, 719.

Chen,W.;Wu,Y.- S. (1990) - *Phys. Review* **D 41**,695.

Ciufolini, I.; Pavlis, E. (2004) - *Letters to Nature*, **431,** 958.

Ciufolini, I. (2005) - Los Alamos Archives http://arxiv.org/abs/gr-qc/0412001 v3

Ciufolini, I.; Wheeler, J. A. (1995) - *Gravitation and Inertia*, Princeton Univ. Press, Princeton. See especially page 82, # 72 to 83.

Coleman, S. (1988) - *Nuclear Physics B* **310**, 643.

Cooperstock,F.I.; (2008) - *General Relativity Dynamics,* World Scientific, Singapore.

Cooperstock,F.I.;Israelit,M. (1995) - *Foundations of Physics,* **25**, 631.

Cooperstock, F.I.; Rosen, N. (1989) - *International Journal of Theoretical Physics,* **28**, 423 - 440.

Cooperstock, F.I.; Tieu, S. (2005) - submitted to Astrophysical Journal. See also Los Alamos Archives http://arxiv.org/abs/astro-ph/0507619.

Cusa, N. (1954)- *Of Learned Ignorance* , Routledge and Kegan Paul , London (see Book II, chapter XI).

Dicke, R.H. (1959) - *Science* **129**, 621.

Dicke, R.H. (1962). Physical Review, **125**, 2163.

Dicke, R.H. (1964) - in *Gravitation and Relativity,* W.A.Benjamin Inc. New York, p1; p. 121.

Dicke, R.H. (1967) - *Physics Today* **20**, 55.

Dicke, R.H. et al. (1965) - *Ap. J.* **142**, 414.

D'Inverno, R. (1992) - *Introducing Einstein's Relativity.* Clarendon Press, Oxford.

Dirac, P.A.M. (1938) - *Proceedings of the Royal Society* **165** A, 199.

Dirac, P.A.M. (1975) - *General Relativity ,* Wiley , New York.

Dolgov, A.D. (1983) - in *The Very Early Universe.* Edited by G.W. Gibbons, S.W. Hawking, and S.T.C. Siklos. Cambridge University Press, Cambridge, p.449.

Dolgov, A.D. (1997) - Los Alamos Archives http://arxiv.org/abs/Astro-ph/9708045.

Dolgov, A.D. (1997) *Physical Review D* **55**, 5881.

Eddington, A.S. (1935). *New Pathways in Science.* Cambridge University Press, Cambridge.

Einstein, A. (1923) - in *Principle of Relativity.* Reprinted by Dover, p.111; p. 177.

Einstein, A. (1921) - *L'Éther et la Théorie de la Relativité*, Gauthiers-Villars, Paris.

Endo, M.; Fukui, T. (1977) - *G.R.G.* **8**, 833.

Eötvös, R. ; Pekar, D. ; Fekete, E. (1922) - Ann. der Phys. **68**, 11.

Falco, E. E., Kochanek, C. S., Muñoz, J. A. (1998), *Astrophys. J.* **494**, 47.

Fardon, R.; Nelson A.E.; Weiner, N. (2003) - Los Alamos Archives http://arxiv.org/abs/astro-ph/0309800 v2.

Fermilab, (2000). News Media Contact. Fermilab, 00-12, July 20, 2000.

Feynman, R. P. (1962-3)- *Lectures on Gravitation*, Addison-Wesley, Reading.

Franklin, A. (2000) - *The Road to the Neutrino*, Physics Today, **53**, N. 2, 22 .

Fujii, Y. and Nishioka, T.(1991) - *Physics Letters B* **254**, 347.

Frieman, J.A.; et al. (1995) - *Physical Review Lett.* **75**, 2077.

Grischuk, L.P (1975)– Sov.Phys. JETP **40**,409

Grischuk, L.P (1975 b)–Lett. Nuovo Cimento **12**, (2) 60.

Grischuk, L.P (1977)– Ann. N.Y Acad. Sci **302**, 439.

Grøn, Ø. (1986). American Journal of Physics, **54**, 46.

Goldstein, H (1980) - *Classical Mechanics,* 2nd. Edtn., Addison Wesley, N.Y.

Gomide, F.M. (1956) - An. Ac. Bras. Ci. **23**, 179.

Gomide, F.M. (1963) - Nuovo Cimento **30**, 672.

Gomide, F.M. (1965) - *An. Ac. Bras. Ci.* **37**, 425.

Gomide, F.M. (1966) - *Nuovo Cimento* **41**, 156.

Gomide, F.M. (1967) - *An. Ac. Bras. Ci.* **39**, 405.

Gomide, F.M. (1972) - *Nuovo Cimento* **12 B**, 11.

Gomide, F.M. (1973) - *Rev. Bras. Fis.* **3**, 3.

Gomide, F.M. (1976) - *Lett. Nuovo Cimento* **15**, 515.

Gomide, F.M. (1980) - *Lett. Nuovo Cimento* **29**, 399.

Gomide, F.M. (1985) - *Rev. Bras. Fis.* **15**,388.

Gomide, F.M.;Berman, M.S. (1988) - *Introduction to Relativistic Cosmology.* (in Portuguese), McGraw-Hill, São Paulo, 2nd ed. (first edition in 1987).

Gomide, F.M.;Berman, M.S.;Garcia, R.L. (1986) - *Rev. Mexicana Astron. Astrofis.* **12**, 46.

Gomide, F.M.;Uehara, M. (1975) - *Prog. Theoretical Physics* **53**, 1365.

Gomide, F.M.;Uehara, M. (1977) - *Rev. Bras. Fis.* **7**, 429.

Gomide, F.M.;Uehara, M. (1978) - *Rev. Bras. Fis.* **8,** 376.

Gomide, F.M.;Uehara, M. (1981) - *Astronomy and Astrophysics* **95**, 362.

Gomide, F.M.;Uehara, M. (1985) - *Ciência e Cultura* **37**, 83.

Gomide, F.M.;Uehara, M. (1985) - *Rev. Bras. Fis.* **15**, 388.

Grischuk, L.P (1975)– Sov.Phys. JETP **40**,409.

Grischuk, L.P (1975 b)–Lett. Nuovo Cimento **12**, (2) 60.

Grischuk, L.P (1977)– Ann. N.Y Acad. Sci **302**, 439.

Grøn, Ø. (1986). American Journal of Physics, **54**, 46.

Guth, A. (1981) - *Physical Review D* **23**, 347.

Halliday, D., Resnick, R., and Walker, J. (2005) *Fundamentals of Physics. 7th.* Edition. Wiley, New York. p. 339. Formula (13-21); etc.

Halverson, N. W. et al. (2002), *Astrophys. J.*, **571**, 604.

Hawking, S.W.(1975) - *Communications of Mathematical Physics* **43**, 199.

Hawking, S.W. (1984) - *Physics Letters B* **134**, 403.

Hawking, S.W. (1993) - *Hawking on the Big-Bang and Black Holes* , World Scientific, Singapore.

Hawking, S. (1996) *The Illustrated A Brief History of Time*, Bantam Books, New York. (1988 earlier non-illustrated edition)

Hawking, S. (2001) - *The Universe in a Nutshell*, Bantam, New York.

Hawking, S. (2003) - *The Illustrated Theory of Everything*, Phoenix Books, Beverly Hills.

Hawking, S.; Mlodinow, L. (2010) - *The Grand Design* , Bantam, New York.

Horvat, R. (2005) - Los Alamos Archives http://arxiv.org/abs/astro-ph/0505507 v2.

Hoyle, F. (1948). MNRAS, **108**, 372.

Iorio, L. ed. (2007) - *The Measurement of Gravitomagnetism* - Nova Science Publishers. New York.

Iorio, L. (2010) - JCAP **08**, 030.

Islam, J.N. (1985) - *Rotating Fields in General Relativity,* CUP, Cambridge.

Kaplan, D.B. (2004) - Physical Review Letters **93**, 091801.

Katz,J.;Ori, A.(1990) *Class. Quantum Grav.***7**,787.

Kenyon, I.R. (1990) - "*General Relativity*". Oxford U.P., Oxford.

Kolb, E. W.; Turner, M. S. (1990) - *The Early Universe*, Addison Wesley, Redwood City,.

Koyré, A.(1962)- *Du Monde Clos à L' Univers Infini* , Presses Universitaires de France

Kramer, D.; Stephani, H.; MacCallum, M.; Heret, E. (1980). *Exact Solutions of Einstein's field equations.* Cambridge U.P., Cambridge.

La,D; Steinhardt,P.J (1989)–Phys. Rev. Lett **62**, 376.

Landsberg, P.T. (1983) - *Private Communication.*

Lederman, L. (1989) - *Observations of Particle Physics from Two Neutrinos to the Standard Model,* Science, **224**, 664 .

Lee, A. T. et al. (2001) - *Astrophys. J. Lett.* **561**, L1.

Lemaître, G. (1930) - MNRAS, **90**, 668.

Lichtenegger, H.; Mashhoon, B. (2007) - *Mach's Principle* in *The Measurement of Gravitomagnetism* , L. Iorio, editor, Nova Science Publishers, New York.

Liddle, A.R.; Wands, D. (1992). Physical Review, **D45**, 2665.

Lima, J.A.S.; Carvalho, J.C. (1994) - *GRG* **26**, 909.

Lima, J.A.S.; Maia, J.M.F. (1994) - *Physical Review D* **49**, 5597.

Lima, J.A.S.; Trodden, M. (1996) - *Physical Review D* **53**, 4280.

Lima, J.A.S. (1996) - *Physical Review* **D 54**, 2571.

Linde, A.D. (1988) - *Physics Letters B* **200**, 272.

Linde, A. (1990). *Particle Physics and Inflationary Cosmology.* Harwood Acad. Press, N.Y.

Lynden-Bell,D.;Katz, J. (1985) - *M.N.R.A.S.* **213**,21.

Mach, E. (1912) - *Die Mechanik in Ihrer ...* Brockhaus, Leipzig.

Maia, M.D.; Silva, G.S. (1994) - *Physical Review* **D 50**, 7233.

Mashhoon, B. (2007) - *Gravitoelectromagnetism,* in *The Measurement of Gravitomagnetism,* ed. L. Iorio, Nova Science Publishers, New York.

Masiero,A.; Vempati,S.K.; Vives,O. (2004) - *Massive Neutrinos and Flavour Violation,* CERN-PH-TH/2004-142.

McCrea, W.H.(1934-35) - *Zs. für Ap.* **9**, 290.

McCrea, W.H.(1951) - *Proc. Royal Society* **206 A**, 563.

McIntosh, C.B.G. (1973). Physics Letters **A43**, 35.

Milgrom, M. (1983) - Astrophysical Journal, **270**, 365-370.

Misner, C.W.; Thorne, K.S.; Wheeler, J.A. (1973) - see MTW below.

Mitra, A. (2006) - Chapter 1 in *Focus on Black Hole Research.* Ed. by Paul V. Kreitler, Nova Science, New York.

Moffat, J.W. (1995) - *Physics Letters B* **357**, 526.

Morganstern, R.E. (1971). Physical Review, **D4**, 278.

"MTW" - Misner, C.W.; Thorne, K.S.; Wheeler, J.A. (1973) *Gravitation* ,Freeman, San Francisco.

Murphy, M.T. et al (2001)– Mon. Not. Royal Astron. Soc, 000, 1-17.

Narlikar, J. (1983). *Introduction to Cosmology*, Bartlett, Boston.

Narlikar, J. (1993). *Introduction to Cosmology*, 2nd. edition (worse than the first one), Cambridge University Press, Cambridge.

Netterfield et al. (2002), *Astrophys. J.* **568**, 38;

Newman, E.T.; et al (1965). Jrn. Math. Phys. **6**, 918.

Ni, W.-T. (2008) - Progress Theor. Phys. Suppl. **172**, 49-60.

Ni, W.-T. (2009) - International Journal of Modern Physcis, **A24**, 3493-3500.

Novello, M. (1980). *Cosmologia Relativista* in *II Escola de Cosmologia e Gravitação*, ed by M.Novello, CBPF, Rio de Janeiro.

Nordtvedt, K. (1970) Astrophys. J, **161**, 1059.

North, J.D. (1965) - *The Measure of the Universe - A History of Modern Cosmology,* Clarendon Press, Oxford.

Ohanian, H.C. (1985) *Physics*, Norton, New York.

O'Hanlon, J.; Tupper, B.O.J. (1972). Nuovo Cimento, **7B,** 305.

Overduin, J.M.; Cooperstock, F.I. (1998) - *Physical Review* **D 58**, 043506.

Özer, M;Taha,M. O .(1986) - *Phys. Lett. B* **171**,363.

Özer, M.;Taha,M. O. (1987) - *Nucl. Phys.B* **287**,776.

Padmanabhan, T. (2002) - Los Alamos Archives http://arxiv.org/abs/hep-th/0212290.

Papapetrou, A.(1974) - *Lectures on General Relativity.* D. Reidel Publishing Company, Boston.

Pathria, R.K. (1972) - Nature **240**, 298.

Peccei, R.D; Solà, J.; and Wetterich, C. (1987) - *Physics Letters* B **195**, 183.

Peebles, P.J.E.; Ratra, B. (1988) - *Astrophysical Journal Lett. Ed.* **325**, L17.

Peratt, A.L. (1990) - The Sciences, N.Y. Academy of Sciences, No. 1, 24.

Perlmutter, S. et al. (1999) - *Astrophys. J.*, **517**, 565.

Pimentel, L.O. (1989) - Astrophysics and Space Science, **112**, 175.

Physics Today (2004) -*Neutrino Oscillation Has Now Been Seen...* , Physics Today, **57**, N. 7, 11, (2004).

Pound, R.V.; Rebka, G.A. (Jr) (1960) - *Physical Review Letters,* **4**, 337.

Purcell, E.M. (1985) - *Electricity and Magnetism*, 2nd. edition, Berkeley Physics Course v2, McGraw-Hill, New York.

Raychaudhuri, A.K. (1975). Physical Review. **53,**1360.

Raychaudhuri, A.K (1979) - *Theoretical Cosmology*, Clarandon Press, Oxford.

Raychaudhuri, A.K.; Banerji, S.; Banerjee, A. (1992). *General Relativity, Astrophysics, and Cosmology,* Springer-Verlag, New York.

Riess, A. G. et al. (1998), *Astron. J.* **116**, 1009.

Rievers, B.; Lamerzahl, C. (2011) - *High precision thermal modeling of complex systems with application to the flyby and Pioneer anomaly* , Los Alamos Archives: arXiv:1104.3985

Rosen, N.(1994) - *Gen. Rel. and Grav.* **26**, 319.

Rowan-Robinson, M. (1981) - *Cosmology*, second edition. Oxford University Press, Oxford. See page 72.

Sabbata, V.de; Sivaran, C. (1994) - *Spin and Torsion in Gravitation* , World Scientific, Singapore.

Sabbata, V.de; Gasperini, M. (1979) - Lettere al Nuovo Cimento **25**, 489.

Salim, J.M.; Waga, I. (1993) - *Classical and Quantum Gravity* **10**, 1767.

Sahni,V.;Starobinski,A. (2000)-*Intl. J.Mod. Phys.* **D9**, 373. Also http://arXiv.org (:astro-ph/9904398 v2).

Sciama, D.N. (1953) - *MNRAS* **113**, 34.

Science, (1992). *Are Neutrinos' Mass Hunters Pursuing a Chimera?*, Science, **256**, 731, (1992).

Science (1992 b). *New Results Yeld no Culprit for Missing Neutrinos*, Science, **256**, 1512, (1992).

Science (1999). *Search for Neutrino Mass...* , Science, **283**, 928, (1999).

Schwarzschild, B. (1992) - Physics Today, **45**, N. 8, 17 .

Schwarzschild, B. (1998) - Physics Today, **51**, N. 8 .

Schwarzschild, B. (2001). Physics Today, **54** (7), 16.

Schweber, S. (2002) - *Enrico Fermi and Quantum Electrodynamics*, Physics Today, **55**, N. 6, 31 .

Shabad, A. E.; Usov, V.V. (2006) - Physical Review Lett. **96**, 180401.

Shojaie, H. (2010) - Los Alamos Archives, arxiv:1005.2261

Shojaie, H.; Farhoudi, M. (2004) - Los Alamos Archives, arxiv:gr-qc/0407096

Shojaie, H.; Farhoudi, M. (2004) - Los Alamos Archives, arxiv:gr-qc/0406027

Silveira, V.; Waga, I. (1994) - *Physical Review D* **50**, 4890.

Silveira, V.; Waga, I. (1997) - *Physical Review D* **56**, 4625.

Stephani, H. (1990) – General Relativity, 2nd edition, CUP, Cambridge.

Su, S.C.; Chu, M.C (2009) - Ap. J. **703**, 354.

Susskind, L.; Lindesay, J. (2005) - *An Introduction to Black Holes, Information and the String Theory Revolution - The Holographic Universe*, World Scientific, Singapore.

Synge, J.L. (1965) - *Relativity - The Special Theory*, 2nd. edition, North Holland, Amsterdan.

Synge, J.L.; Schild, A. (1969) - *Tensor Calculus*, reprinted in 1978, with corrections, Dover, New York.

Taylor, J.H. ; Weisberg, J.M. (1982) - *Astrophysics Journal,* **253**, 908.

Taylor, J.H. ; Weisberg, J.M. (1989) - *Astrophysics Journal,* **345**, 434.

Tolman, R.C.(1934) - *Relativity, Thermodynamics and Cosmology* , OUP. Reprinted by Dover, New York (1987).

Torres, L.F.B.; Waga, I. (1996) - *MNRAS* **279**, 712.

Turyshev, S.G.; Toth, V.T. (2010) - *The Pioneer Anomaly,* Los Alamos Archives arXiv:1001.3686 [gr-qc]

Uehara, K.; Kim, C.W. (1982) - *Physical Review* **D26**, 2575.

(Unknown, 1992) - Science **256**, 731.

(Unknown, 1992) - Science **256,** 1512.

(Unknown, 1999) - Science **283**, 928.

(Unknown, 2000) - *News Media Contact,* Fermilab 00-12. July, 20.

(Unknown, 2004) - Physics Today **57**, N.7, 11.

Virbhadra, K.S. (1990) - *Phys. Rev.* **D41**, 1086; **D42**, 2919; **D42**, 1066. See also Aguirregabiria, J.M. et al. (1996) - *Gen. Rel. and Grav.* **28**, 1393.

Waga, I. (1993) - *Ap. J.* **414**, 436.

Wagoner, R.V. (1970)- Phys. Rev. D**1** , 3209.

Webb,J.K. , et al. (1999)–Phys. Rev. Lett. **82**, 884.

Webb,J.K; et al.(2001)– Phys Rev. Lett. **87,** 091301.

Weinberg, S. (1972) - *Gravitation and Cosmology*, Wiley, New York.

Weinberg, S. (1989) - *Reviews of Modern Physics,* **61**, 1.

Weinberg, S. (2008) - *Cosmology*, OUP, Oxford.

Wrede, R.C.; Spiegel, M. (2002) - *Advanced Calculus*, Second Edition, McGraw-Hill, New York.

Wesson, P.S. (1999) - *Space-Time-Matter (Modern Kaluza-Klein Theory)*, World Scientific, Singapore.

Wesson, P.S. (2006) - *Five Dimensional Physics*, World Scientific, Singapore.

Weyl, H. (1950) - *Space-time-matter*, Dover, N.Y.

Whitrow, G. (1946) - *Nature* **158**,165.

Whitrow, G.; Randall, D. (1951)-*MNRAS* **111**,455.

Whittacker, E.(1935) - *Proc. Royal Society* **149 A**, 384.

Whittacker, J.M. (1966) - Nature. **209**, 491.

Will, C.M. (1984) Phys. Rep **113**, 345.

Wrede, R.C.; Spiegel, M. (2002) - *Advanced Calculus*, Second Edition, McGraw-Hill, New York.

Zel'dovich, Ya. B. (1967) Pi'sma *JETP* .**6**, 887.

Zel'dovich, Ya. B. (1967 b) *JETP Lett.***6**, 316.

Zel'dovich, Ya. B.(1968) - *Soviet Physics Usp.* **11**, 381.

Zel'dovich, Ya. B. (1981) *Sov. Phys. Usp.* **24**, 216.

INDEX

S

T

U

V

W

Y